PRAISE FOR

THE INDUSTRIES OF THE FUTURE

"A fascinating vision of the future of industry. *The Industries of the Future* reads like a portable TED conference at which you have been seated next to the smartest guy in the room. The book is filled with glimpses of cutting-edge biotech research, statecraft, and entrepreneurship. Ross writes engagingly, and the book should be compelling whether you follow these fields closely or you still think of Honda as a car rather than a robotics company."

—*Forbes*

"In a world growing more chaotic, Alec Ross is one of those very rare people who can see patterns in the chaos and provide guidance for the road forward. He has an unusual diversity of expertise that allows him to apply multiple lenses to the world's challenges and dream up the kind of innovative solutions that are changing the world."

—Eric Schmidt, executive chairman, Alphabet, and author of *The New Digital Age*

"Brilliant, captivating . . . Alec Ross combines an extraordinary understanding of future trends with an equally extraordinary ability to describe those trends and explain what they mean for the world in the decades ahead."

—General (Ret.) David H. Petraeus, chairman, KKR Global Institute; former commander of the surge in Iraq; and director of the CIA

"A riveting and mind-bending book. If you want to know how to survive and thrive in the fast-paced world of today and how to anticipate the opportunities of tomorrow . . . this is a good place to start."

—*New York Journal of Books*

"An engaging, clear-eyed look at the benefits and challenges of the coming wave of global innovation. Alec Ross, with his years of passionate work in the public and private sectors, is uniquely positioned to understand and explain where we're coming from and where we're going."

—Arianna Huffington

"A lucid and informed guide, even on the most technical issues."

—*Financial Times*

"Inspiring and sobering at the same time."

—*Fast Company*

"[*The Industries of the Future*] will take its place alongside other classic tech-and-society books, like Tim Wu's *The Master Switch* and Jonathan Zittrain's *The Future of the Internet*. Ross . . . reveals not just where industries are heading but where entire societies may end up as a result."

—*Medium*

"A must-read for the rising generation and their concerned parents and educators."

—Wendy Kopp, founder, Teach for America, and CEO, Teach for All

"Anyone who wants to understand the key forces that are shaping our economic, political, and social futures will benefit hugely from Ross's insights."

—Reid Hoffman, founder and chairman, LinkedIn

"Astute and enlightening."

—*Publishers Weekly*

"Articulate, accessible, and engaging . . . Whether you are a business executive considering new investments and navigating emerging markets or an advocate for human rights issues, this book is your guide to the geopolitical, cultural, and technological contexts that will impact your work."

—*Global Daily*

"Wide-ranging and smoothly written . . . Incisively covering economics, politics, cyberwarfare, genomics, and the complexities of Big Data—among many other things—in language for the layperson interested in what tomorrow could bring."

—*Booklist*

"Discerning."

—*Kirkus Reviews*

"Well worth reading."

—*The Washington Times*

"An important read for everyone from businesspeople to parents to teachers."

—*TechRepublic*

"The future is already hitting us, and Ross shows how it can be exciting rather than frightening."

—Walter Isaacson, author of *Steve Jobs* and *The Innovators*

THE INDUSTRIES OF THE FUTURE

ALEC ROSS

SIMON & SCHUSTER PAPERBACKS

NEW YORK LONDON TORONTO SYDNEY NEW DELHI

Simon & Schuster Paperbacks
An Imprint of Simon & Schuster, Inc.
1230 Avenue of the Americas
New York, NY 10020

First Simon & Schuster trade paperback edition February 2017

SIMON & SCHUSTER PAPERBACKS and colophon are
registered trademarks of Simon & Schuster, Inc.

For information about special discounts for bulk purchases,
please contact Simon & Schuster Special Sales
at 1-866-506-1949 or business@simonandschuster.com.

The Simon & Schuster Speakers Bureau can bring authors to
your live event. For more information or to book an event, contact
the Simon & Schuster Speakers Bureau at 1-866-248-3049
or visit our website at www.simonspeakers.com.

Interior design by Ruth Lee-Mui

Manufactured in the United States of America

10   9   8   7   6   5

The Library of Congress has cataloged the hardcover edition as follows:
Ross, Alec, 1971– author.
The industries of the future / Alec Ross.
pages cm
Includes bibliographical references and index.
1. Technological innovations—Economic aspects. 2. Industries—Technological innovations.
3. Research, Industrial. I. Title.
HC79.T4R677 2016
338'.064—dc23
2015028769

ISBN 978-1-4767-5365-2
ISBN 978-1-4767-5366-9 (pbk)
ISBN 978-1-4767-5367-6 (ebook)

To my wife, Felicity, who keeps our family together and grounded when I am in the air, far away, far too often.

CONTENTS

THE INDUSTRIES OF THE FUTURE

INTRODUCTION

Adapt or perish, now as ever, is nature's inexorable imperative.

—H. G. Wells, *A Short History of the World* (1922)

THE WRONG SIDE OF GLOBALIZATION

It's 3:00 a.m., and I'm mopping up whisky-smelling puke after a country music concert in Charleston, West Virginia.

It's the summer of 1991, just after my freshman year of college. While most of my friends from Northwestern University are off doing fancy internships at law firms, congressional offices, and investment banks in New York or Washington, I am one of six guys on the after-concert janitorial crew at the Charleston Civic Center, which seats 13,000 people.

Working the midnight shift is worse than jet lag. You have to decide if you want your work to be the beginning of your day or the end of your day. I would wake up at 10:00 p.m., eat "breakfast," work from midnight to 8:00 a.m., and then go to bed around 3:00 p.m.

The other five guys on the crew were a tough bunch. They were good guys but beaten down. One carried a pint bottle of vodka in his back pocket, which was done by "lunch" at 3:00 a.m. A scraggly red-head from the hollows, the valleys that run between West Virginia's hills, was sort of near my age. The others were in their 40s and 50s, at what should have been the peak of their wage-earning potential.

The way country music concerts work in West Virginia is people drink way too much. Our job was to clean up the result. The six of us canvassed the arena with enormous jugs of fluorescent-blue chemicals, which, when poured on the concrete floor, would just sizzle.

The last wave of innovation and globalization produced winners and losers. One group of winners were the investors, entrepreneurs, and high-skilled laborers that congregated around fast-growing markets and new inventions. Another class of winners were the more than 1 billion people who moved from poverty into the middle class in developing countries because their relatively low-cost labor was an advantage once their countries opened up and became part of a global economy. The losers were people who lived in high-cost labor markets like the United States and Europe whose skills could not keep up with the pace of technological change and globalizing markets. The guys I mopped with on the midnight shift were the losers in large part because the job they could have gotten in a coal mine years before had been replaced by a machine, and whatever job they could have gotten in a factory from the 1940s to the 1980s had moved to Mexico or India. For these men, being a midnight janitor was just not the summer job it was to me; it was one of the only options left.

Growing up, I thought that life in West Virginia was representative of life everywhere. You were doing your best to manage a slow descent. But the phenomenon I was witnessing in West Virginia really made sense to me only as I traveled the world and saw other regions rising as West Virginia was falling.

Twenty years after pushing a mop on the midnight shift, I've now seen the world and been exposed to the highest levels of leadership in the biggest technology companies and governments around the world.

I've served as Secretary of State Hillary Clinton's senior advisor for innovation, a position she created for me just as she became known as Madame Secretary. Before going to work for Clinton, I served as the convener for technology and media policy on the Obama campaign that beat her in the 2008 presidential primary and had spent eight years helping run a successful, technology-based social venture that I cofounded. My job at the State Department was to modernize the practice of diplomacy and bring new tools and approaches to addressing foreign policy challenges. Clinton recruited me to bring a little innovation mojo to the tradition-bound State Department. We had a lot of success, and at the time that she and I left in 2013, we were ranked as having the most innovation-friendly culture of any cabinet-level department in the federal government. We developed successful programs to address nasty challenges in places as varied as the Congo, Haiti, and the cartel-controlled border towns in northern Mexico. In the background of all this was the role I played building a bridge between America's innovators and America's diplomatic agenda.

In this time, I spent much of my life on the road. I saw a lot of the world before and after my time in government, but the 1,435 days I spent working for Hillary Clinton gave me a particularly intense, close-up view of the forces shaping the world. I traveled to dozens and dozens of countries, logging more than half a million miles, the equivalent of a round trip to the moon with a side trip to Australia.

I saw next-generation robotics in South Korea, banking tools being developed in parts of Africa where there were no banks, laser technology used to increase agricultural yields in New Zealand, and university students in Ukraine turning sign language into the spoken word.

I have had the chance to see many of the technologies that await us in the coming years, but I still often think back to that stint as a midnight janitor and the men I met there. The time I spent gaining a global perspective on the forces shaping our world helped me understand exactly why life had grown so rough in my home in the hills and why life was getting so much better for most of the rest of the world.

The world in which I grew up, the old industrial economy, was radically transformed by the last wave of innovation. The story is by now well worn: technology, automation, globalization.

While I was in college in the early 1990s, the process of globalization accelerated further, bringing to an end many of the political and economic systems that defined yesterday's economies. The Soviet Union and its satellite states failed. India began a series of economic reforms to liberalize its economy, eventually bringing more than a billion people onto the global economic playing field. China reversed its economic model, creating a new form of hybrid capitalism and pulling more than half a billion people out of poverty.

The European Union was created. The North American Free Trade Agreement (NAFTA) came into effect, integrating the United States, Canada, and Mexico into what is now the world's largest free trade zone. Apartheid ended and Nelson Mandela was elected president of South Africa.

While I was in college, the world was also newly coming online. The World Wide Web was launched to the public, along with the web browser, the search engine, and e-commerce. Amazon was incorporated while I was driving to a training site for my first job out of college.

At the time, these political and technological shifts did not seem as important to me as they do now, but the changes that took place while I was growing up in West Virginia and that accelerated with the rise of the Internet have made the lives we lived even just 20 years ago seem like distant history.

Those people in my hometown with worse job security than their parents are still living a better life when you measure it up against what their money can buy today that it could not decades ago, including more and better communications and entertainment, healthier food, and safer cars and medical advances that keep them alive longer. Yet they've been through a raft of changes, both positive and negative. And all this change will pale in comparison to what is going to come in the next wave of innovation as it hits all 196 countries on the planet.

The coming era of globalization will unleash a wave of technological, economic, and sociological change as consequential as the changes that shook my hometown in the 20th century and the changes brought on by the Internet and digitization as I was leaving college 20 years ago.

In business areas as far afield as life sciences, finance, warfare, and agriculture, if you can imagine an advance, somebody is already working on how to develop and commercialize it.

The places where innovation gets commercialized are expanding too. In the United States, breakthroughs are coming not only from Silicon Valley, from the Route 128 corridor around Boston, or from North Carolina's research triangle. They are beginning to come out of Utah, Minnesota, and the Washington, D.C., suburbs in Virginia and Maryland. The breakthroughs will not be exclusively American, either.

After years of growth rooted in low-cost labor, there are promising signs of innovation coming from the 3 billion people who live in Indonesia, Brazil, India, and China. Latin American countries with a face to the Pacific, including Chile, Peru, Colombia, and Mexico, appear to have figured out how to position themselves in the global economy. The highest-skilled labor markets in Europe are producing start-ups that make Silicon Valley green with envy, and in tiny Estonia, "the little country that could," the entire economy seems to be an e-economy.

Innovation is likewise transforming Africa, where even in the refugee camps of the Congo, technologies as simple as the cell phone are connecting people to information and each other as never before. Africa's entrepreneurs are now changing the face of the continent, fueling development and creating a new class of globally competitive businesses.

Everywhere, newly empowered citizens and networks of citizens are challenging the established order in ways never before imaginable— from building new business models to challenging old autocracies.

The near future will see robot suits that allow paraplegics to walk, designer drugs that melt away certain forms of cancer, and computer code being used as both an international currency and a weapon to destroy physical infrastructure halfway around the world.

This book examines these breakthroughs, but it is not simply a hosanna to the benefits of innovation. Advances and wealth creation will not accrue evenly. Many people will gain. Some people will gain hugely. But many will also be displaced. Unlike the previous wave of digital-led globalization and innovation, which drew enormous numbers of people out of poverty in low-cost labor markets, the next wave will challenge middle classes across the globe, threatening to return many to poverty. The previous wave saw entire countries and societies lifted up economically. The next wave will take frontier economies and bring them into the economic mainstream while challenging the middle classes in the most developed economies.

Across large swaths of the globe, people feel newly under siege by rising inequality and unwelcome disruption. A pervasive sense that it is becoming harder to find your place in the world or get ahead is rattling many societies.

Innovation brings both promise and peril. The same forces that are unleashing unparalleled advances in wealth and welfare may also allow a hacker to steal your identity or hack your home. A computer that can speed up analysis of legal documents can also shrink the number of lawyers in the workforce. Social networks can open doors to form new connections or create new forms of social anxiety. The digitization of payments can facilitate commerce or allow for new forms of fraud.

When I was a college student at the dawn of the Internet revolution, I did not have the slightest inkling of the future that lay ahead. I wish I had been able to read a book back then that took a good stab at what was next. Certainly no one is omniscient, but I have been fortunate enough to gain a glimpse of what lies around the next corner.

This book is about the next economy. It is written for everyone who wants to know how the next wave of innovation and globalization will affect our countries, our societies, and ourselves.

GROWING UP IN THE OLD ECONOMY

To understand where globalization is going in the future, you have to understand where it's coming from. I grew up in Charleston, West Virginia, a city whose history reflects America's centuries-long rise as an economic powerhouse from the grime-covered mines that helped fuel its growth. West Virginia was built on coal, much the same way that Pittsburgh was built on steel and Detroit was built on cars. Indeed, it was West Virginia's connections through coal to the industrial North that led it to secede from Virginia and the more agricultural South when the American Civil War broke out.

West Virginia's position mirrored that of other mining centers connected to the Industrial Revolution's first manufacturing bases. In the United Kingdom, Midlands cities like Manchester and Leeds became the industrial base. London provided the finance. Coal came from Wales. In Germany, the Ruhr region near the Rhine River valley became a manufacturing center. Coal came from eastern Germany and Poland.

Today coastal China, particularly the areas around Shenzhen and Shanghai, has become the world's factory. Its coal comes from western China and Australia. Similarly, mining regions in India's northeastern peninsular belt, Turkey's Anatolia region, and Brazil's Santa Catarina region supply the industrial bases of their emerging economies and other economies around the world. In each region, mining offers a stepping-stone toward greater economic ties and opportunities—at least for a time.

Building on its coal boom, West Virginia developed complementary industries that cemented its position as an industrial supply center and eventually presaged its decline. In the early 20th century, Charleston underwent its second boom: chemicals. The Union

Carbide Corporation established the world's first petrochemical plant in West Virginia in 1920.

With America's entry into World War II, massive amounts of synthetic rubber were needed to meet wartime demands. Union Carbide, which became the largest employer in West Virginia and one of the top ten chemical companies in the world, launched a period of growth that continued well after the war. Between 1946 and 1982, its revenues increased from about $415 million to more than $10 billion. During that time, it employed as many as 80,000 people globally, roughly 12,000 of them in West Virginia. And as the company continued to grow, so did Charleston. By 1960, its population had grown to 86,000 from 68,000 before the war.

When I was in school, a big percentage of my classmates were children of chemical company engineers. Their families were often the most worldly, arriving from top universities around the country and the globe. And for over a century, the old-economy industrial fields of West Virginia—coal, chemicals, and plastics—were stable, reliable career choices.

My family came to West Virginia when my grandfather, Ray DePaulo, moved from the coal camps of Colorado during the Great Depression. His high school had closed down due to lack of funds, so they just handed out diplomas to everyone, including my then 13-year-old grandfather. Luckily for him, that was still a time when you could make a living with just a high school diploma.

Once in West Virginia, my grandfather became what we'd now glowingly call an entrepreneur. He went door-to-door selling telephones at a time when people were getting phones in their homes for the first time. He ran a garage, a golf course, a restaurant, a bakery, a parking lot, and a house-cleaning business, much of it out of a stall used by car salesmen.

My grandfather understood one of the curious conundrums of globalization: exposure creates not only opportunity but competition, and it can make us question and eventually lose our standing in the world. West Virginia, like so many other industrial centers in America, was just realizing its high point during my grandfather's lifetime. But its shortcomings were about to be exposed by new competition from new markets and new machines.

I remember, as a kid, about halfway into the drive between Charleston and my father's law office was a town named Nitro. As soon as we passed Nitro, my brother, sister, and I would all start squirming and squealing in our seats, repulsed by the stench from the chemical factories that surrounded us.

My mother, in the driver's seat, would just breathe it in and matter-of-factly say, "That's the smell of money." She identified this horrid smell with jobs—and potential clients for my dad.

In the old economy, this was the smell of money not just in Nitro but in places like Gary, Indiana; Newark, New Jersey; and Baton Rouge, Louisiana. Today that same smell now shrouds industrial sites in China, India, and Mexico even as it still lingers in some of the old industrial centers of America.

The Kanawha River Valley, which runs through Charleston, was known as "Chemical Valley." For almost a century, Chemical Valley hosted the highest concentration of chemical manufacturers in the United States: Union Carbide, DuPont, Monsanto, the Food and Machinery Corporation (FMC), and many, many others.

At night, the valley looked from above like something out of a futuristic film with its towering steel structures of the chemical plants lit up with little lights. The smoke released into the night sky added an orange glow, the whole scene eerily reflected by the river—a river that seemed not to have a single fish or other living animal in it. I never questioned why as I was growing up.

The town of Nitro, 14 miles downriver from Charleston, is named for nitrocellulose, or gunpowder, bringing a literal meaning to the term *boom town*. Nitro was built up during America's mobilization for World War I, when the United States faced a critical shortage in gunpowder production. The US government poured more than $70 million into building the plant, known as Explosive Plant C, as well as the public structures for the surrounding town. The war ended in November 1918, just after the first shipment of explosives went out.

That was not the last time Nitro participated in a war mobilization. In the 1960s, the former Monsanto chemical plant in Nitro manufactured the herbicide Agent Orange, used by the US military to defoliate jungles during the Vietnam War. Dropping this chemical in the Vietnamese jungle harmed the health of more than 1 million Vietnamese and 100,000 American combat veterans, and it produced birth defects in more than 100,000 children.

It was an ugly trade, but for a while, for the local region, my mother was right: the smell of Nitro meant money. The money did not keep coming, however. Ultimately the very industries that built Charleston and West Virginia also led to their decline.

In the case of the mining industry, mechanization wiped out the need for coal mining by hand. To be a coal miner went from carrying a pick and shovel to operating a machine that could do the job of hundreds of humans. For the chemical industry, globalization meant that businesses could locate their plants in places with fewer environmental regulations and less expensive labor. The chemical companies moved their operations to India and Mexico.

Leaving home became the practical option for many Charlestonians. From 1960 to 1990, Charleston lost 40 percent of its population. By 1988, West Virginia's unemployment rate was close to double the national average. There were fewer children of engineers in my classes as their parents were transferred out of the state or out of the country.

Charleston and West Virginia were representative of cities and states all over the world grappling with postindustrial decline and globalization. Handcuffed to their existing resources and manufacturing industries, these areas flourished through years of stable economic growth. But once the boom ended, they were hit hard with rapid capital and population flight. Their manufacturing facilities, once drivers of wealth, today are dormant steel structures on the city skyline. The steel mills of Pittsburgh shuttered. Detroit's population declined from 1.8 million to 700,000 as its car industry faced new competition from Tokyo and Seoul.

American cities were not the only ones to suffer. Manchester, the world's first industrialized city, lost 50,000 jobs in the 1970s. The coalfields in southern Wales went under, with the last one closing in 2008. The port of Marseilles was ravaged by new competition and saw its population decrease by 150,000 people.

By the time I was in college, the only chemicals that provided secure jobs in West Virginia were the ones that sizzled as they hit the floor of the Charleston Civic Center.

THE RIGHT SIDE OF GLOBALIZATION

As West Virginia faced a decades-long economic descent, those same forces of globalization and labor migration were having positive impacts elsewhere. In India and China, together 40 percent of the world's population, the changes have been eye-popping.

In the 30 years from 1982 to 2012, India's poverty rate dropped from 60 percent of the population to 22 percent. Life expectancy surged from 49 years to 66. When I was growing up, India was the country of Mother Teresa and famine. Today it is a country increasingly defined by technology, global services, and a fast-growing middle class.

The changes in China have been even more dramatic, with its poverty rate plummeting over that same period from 84 percent to 13 percent, pulling roughly 600 million Chinese out of poverty. With an economy 25 times larger than it was 30 years ago, China has become the second largest economy in the world after the United States.

What was bad for industrial America and Europe was very good for India, China, and much of the rest of the world. As globalization and innovation challenged the lifestyles of many of those living in industrial cities and states in the West, it bolstered the economic growth of up-and-coming nations. Beyond developing nations, individuals and states all over the world that took advantage of the wave of technological innovation flourished. Our most valued commodities have gone from salt and sugar to chemicals and fuels to data and services. The regions that provide those now lead the global knowledge economy. Twenty-five hundred miles from Charleston, West Virginia, several trillion dollars of wealth was generated in Silicon Valley in addition to products that fundamentally changed the way everyone reading this book lives.

THE INDUSTRIES OF THE FUTURE

The book I know my parents or grandparents wish they had read in the 1960s would have described what globalization was going to do to the world. The book I wish I had read as I left college more than 20 years ago would have told me what the Internet and digitization were going to do to the world. This book explores the industries that will drive the next 20 years of change to our economies and societies. Its chapters are built around key industries of the future—robotics, advanced life sciences, the code-ification of money, cybersecurity, and big data—as well as the geopolitical, cultural, and generational contexts out of which they are emerging. I chose these industries not only because they are important in their own right but because they are also symbolic of larger global trends and symbiotic among each other.

"Here Come the Robots" and the "The Future of the Human Machine," chapters 1 and 2, explore how cutting-edge advances in robotics and the life sciences will change the way we work and live—with enormous, but unequal, impact on our livelihoods and our lives. As robots increasingly occupy the world alongside us, the global economy will undergo a revolution spurred by artificial intelligence and machine learning that could be as consequential for labor forces as the agricultural, industrial, and digital revolutions that preceded it. At the same time, dramatic advances in life sciences will allow people to live longer and healthier lives than ever before—at least those who can afford it. The economic returns in robotics and life sciences will likewise be unevenly distributed between those who are well positioned to create or adopt these new breakthroughs and those who may be left even further behind. In response, societies will need to find new ways to adapt.

"The Code-ification of Money, Markets, and Trust" and "The Weaponization of Code" (chapters 3 and 4) examine how the increasing application of computer code to new areas of the economy—in the virtual and physical worlds—will transform two spheres that are traditionally state monopolies: money and force. Rapid progress often comes with greater instability. The application of code to commerce will provide new opportunities for the proverbial little guy in any part of the world to receive, hold, spend, or transfer money. At the same time, from the vantage point I had in Secretary Clinton's office and the White House Situation Room, I saw the future of an industry that has gone from being a small back-office IT function to one of the fastest-growing and most disruptive industries in the world: the weaponization of code. Together these developments may bring new opportunities, but they will also increase the ability of bad actors to cause systemic damage to the international economy.

"Data: The Raw Material of the Information Age" and "The Geography of Future Markets" (chapters 5 and 6) examine both the expansiveness that big data will allow and the constraints that geopolitics will place on the global marketplace. Whereas land was the raw material of the agricultural age and iron was the raw material of the industrial age, data is the raw material of the information age. The Internet has become an ocean of jumbled, chaotic information, but now there is a way to connect this information and draw actionable business intelligence from it. Big data is transitioning from a tool primarily for targeted advertising to an instrument with profound applications for diverse corporate sectors and for addressing chronic social problems.

At the same time, the industries of the future will both be created within the current geopolitical structure and transform it. In the 20th century, the dominant divide between political systems and markets was along the axis of left versus right. In the 21st century, the dominant divide is between those that have open political and economic models and those that are closed. New competition and political necessity have created a series of hybrid models around the world, and these final two chapters explore what markets will be future sources of sustainable growth and innovation and how business leaders can make informed choices about where to invest their time and resources.

Throughout the book, we explore competitiveness—what it takes for societies, families, and individuals to thrive. Among the world's most innovative countries and businesses there is an emerging cultural consensus on how best to strengthen their most critical resource: their people. And there is no greater indicator of an innovative culture than the empowerment of women. Fully integrating and empowering women economically and politically is the most important step that a country or company can take to strengthen its competitiveness. Societies that do not overcome their negative cultural legacies regarding the treatment of women will founder in the next wave of innovation. The world's most restrictive countries have been absent from the most recent wave of innovation, and they will not be home to industries and businesses of the future without making real changes. Innovation doesn't happen in closed environments, and innovative companies will continue to steer clear of countries with regressive policies on gender.

Last, the book looks forward to explore what interventions we can make in our children's lives to best prepare them for success in a world of increasing change and competition. Parenting is the most important job that a person can have, and our children will grow up to inherit a world that looks much different from our own. We can draw from the wisdom of the innovators profiled in these pages to prepare both ourselves and our children for what's coming in the next economy—for the economy that begins now.

ONE

HERE COME THE ROBOTS

Welcome your new job takers and caregivers. The coming decade
will see societies transform as humans learn to live alongside robots.

Japan is home to the longest-living citizens on earth and the biggest elderly population of any country—and it's not getting any younger. Japan's current life expectancy is 80 years for men and 87 years for women and is expected to rise to 84 and 91, respectively, over the next 45 years. Between 2010 and 2025, the number of Japanese citizens 65 years or older is expected to increase by 7 million. Today, 25 percent of Japan's population is age 65 or older. By 2020, this is projected to increase to 29 percent and reach 39 percent by 2050.

All of those long-living elderly will need caretakers. Yet Japan's low birthrates mean that what once was a staple of Japanese family life—taking care of one's grandparents and great-grandparents—will no longer be a viable model at the scale the nation needs. There will not be enough grandchildren.

With Japan's persistently strict immigration policies curtailing the number of workers in the country, there will not be enough humans around to do the job at all. Japan's Ministry of Health, Labor, and Welfare predicts a need for 4 million eldercare nurses by 2025. Right now there are only 1.49 million in the country. Japan allows only 50,000 work visas annually, and unless something drastic changes, the math does not work.

This labor shortage will hit service-industry jobs like eldercare with ferocity and will be exacerbated because caretakers have a high job turnover rate due to low pay and high rates of work-related injury from lifting patients.

Enter the robots.

Our future caretakers are being developed in a Japanese factory right now. Just as Japanese companies reinvented cars in the 1970s and consumer electronics in the 1980s, they are now reinventing the family. The robots depicted in the movies and cartoons of the 1960s and 1970s will become the reality of the 2020s.

Rival Japanese companies Toyota and Honda are leveraging their expertise in mechanical engineering to invent the next generation of robots. Toyota built a nursing aide named Robina—modeled after Rosie, the cartoon robot nanny and housekeeper in *The Jetsons*—as part of their Partner Robot Family, a line of robots to take care of the world's growing geriatric population. Robina is a "female" robot, 60 kilograms in weight and 1.2 meters tall, that can communicate using words and gestures. She has wide-set eyes, a moptop hairdo, and even a flowing white metallic skirt.

Robina's brother, Humanoid, serves as a multipurpose home assistant. He can do the dishes, take care of your parents when they're sick, and even provide impromptu entertainment: one model plays the trumpet, another the violin. Both versions are doppelgangers for the famous *Star Wars* C-3PO robot, although in gleaming white instead of gold.

In response, Honda has created ASIMO (the Advanced Step in Innovative Mobility robot), a fully functional humanoid that looks like a four-foot-tall astronaut stuck on Earth. ASIMO is sophisticated enough to interpret human emotions, movements, and conversation. Equipped with cameras that function as eyes, ASIMO can follow voice commands, shake hands, and answer questions with a nod or by voice. He even bows to greet others, demonstrating good Japanese manners. For an elderly patient, ASIMO can fulfill a range of tasks, from helping the patient get out of bed to holding a conversation.

Honda is also focusing much of its research and commercialization on robotic limbs and assistance devices that are robotic but not freestanding robots. Its Walking Assist device wraps around the legs and back of people with weakened leg muscles, giving them extra power to move on their own. In the future, expect to see Honda making robotic hands and arms. Its goal is nothing less than helping paraplegics walk and the very frail rediscover the speed and power of their youth.

Numerous other Japanese companies are pushing the big players like Toyota and Honda. Tokai Rubber Industries, in conjunction with the Japanese research institute RIKEN, has unveiled the Robot for Interactive Body Assistance (RIBA), which can pick up and set down humans up to 175 pounds and is designed for patient comfort: it resembles a giant smiling bear and is covered in a soft skin to guard against injury or pain. Similarly, Japanese industrial automation company AIST has created PARO, a robot baby harp seal covered in soft white fur. PARO exhibits many of the same behaviors as a real pet. Designed for those who are too frail to care for a living animal or who live in environments that don't allow pets, such as nursing homes, it enjoys being held, gets angry when hit, and likes to nap. When President Barack Obama met PARO a few years ago on a tour of Japanese robotics innovations, he instinctually reached out and rubbed its head and back. It looks like a cute stuffed animal, but costs $6,000 and is classified by the US government as a class 2 medical device.

Japan already leads the world in robotics, operating 310,000 of the 1.4 million industrial robots in existence across the world. It's turning to eldercare robots in part because it has to and in part because it, uniquely, is in a great position to leverage its advanced industrial technology toward the long assembly line of the human life span. But can robots really take care of humans?

Japan's private and public sectors certainly think so. In 2013, the Japanese government granted $24.6 million to companies focusing on eldercare robotics. Japan's prominent Ministry of Economy, Trade, and Industry chose 24 companies in May 2013 to receive subsidies covering one-half to two-thirds of the R&D costs for nursing care robots. Tasks for these robots include helping the elderly move between rooms; keeping tabs on those likely to wander; and providing entertainment through games, singing, and dancing.

Nevertheless, difficult challenges remain. On the technical side, it remains difficult to design robots capable of intimate activities like bathing patients or brushing their teeth. And most Japanese companies that are developing these robots specialize in industrial motors and electronic automation. They didn't enter the caretaking field with a keen grasp of how to forge an emotional connection, a crucial aspect of eldercare. Even as they improve, some observers— like Sherry Turkle, a professor of the social studies of science and technology at MIT—question whether patients will ever be able to form a true emotional connection with robot caretakers. As Turkle warns, "For the idea of artificial companionship to be our new normal, we have to change ourselves, and in the process we are remaking human values and human connection." If robot nurses catch on, she explains, they may even create a chasm between younger and older generations. "It's not just that older people are supposed to be talking," Turkle argues, referring to the goal of creating robots that can hold conversation, "younger people are supposed to be listening. We are showing very little interest in what our elders have to say. We are building the machines that will literally let their stories fall on deaf ears."

These technical questions (Can a robot brush a person's teeth?) and almost-spiritual doubts (Can, and should, emotional connections be made between humans and robots?) are both valid. Yet robot technology and applicability continue to advance in Japan, and answers to these questions will likely arise there in the near future. With too few caretakers, I expect robots to become a regular part of the Japanese family system.

If the aging nation can pull it off, robot caretakers will be a boon for its economy and will soon make the jump to the global economy, with potentially far-reaching consequences.

Much of the rest of the industrialized world is on the verge of a period of advanced aging that will mirror Japan's own. In Europe, all 28 member states of the European Union have populations that are growing older, and in the decades ahead, the percentage of Europe's population aged 65 and older will grow from 17 percent to 30 percent. China is already entering a period of advanced aging even as it continues to develop. Although its one-child policy is already being phased out, China is now demographically lopsided. Chinese women have on average 1.4 children, well below the replacement rate of 2.1, resulting in too few young people to provide for the elderly. The notable exception is the United States, where immigration policies partially mitigate the effects of an aging population.

As the populations of developed nations continue to age, they create a big market for those Japanese robots. And caretaking robots, alongside robotic limb technology, may simply be the first in a new wave of complex robots entering our everyday lives. Robots will be the rare technology that reaches the mainstream through elderly users first, spreading down as grandma shows off her next cutting-edge gadget for the kids and grandkids.

THE GEO-ROBOTIC LANDSCAPE

The robot landscape will be vastly differentiated by country. Just as wealthier and poorer citizens reside at different technological levels, so do wealthier and poorer countries.

A few countries have already established themselves as leading robot societies. About 70 percent of total robot sales take place in Japan, China, the United States, South Korea, and Germany—known as the "big five" in robotics. Japan, the United States, and Germany dominate the landscape in high-value industrial and medical robots, and South Korea and China are major producers of less expensive consumer-oriented robots. While Japan records the highest number of robot sales, China represents the most rapidly growing market, with sales increasing by 25 percent every year since 2005.

There is quite a gap between the big five and the rest of the world. As both consumers and producers of robots, these countries outpace all others. By way of illustration, the number of industrial robots produced in South Korea, a country of 50 million people, is several times greater than the number produced in South America, Central America, Africa, and India combined, with populations totaling 2.8 billion. Russia is effectively a nonplayer in robotics despite its industrial base. It neither produces nor buys robots to any significant degree, instead maintaining extractive industries (natural gas, oil, iron, nickel) and industrial manufacturing plants that look and function the way they did in the 1970s and 1980s.

The big five's comparative advantage might even accelerate in the future, for these are the same countries that are most likely to incorporate the next generation of robotics into society, work, and home. They will own the name brands in consumer robots, and they'll power the software and networks that enable the robotics ecosystem. When I think about this symbiosis, I think about the Internet in the 1990s. It was not just the consumer-facing Internet companies that were born and based in Silicon Valley; it was also the network equipment makers like Cisco Systems and Juniper Networks. Today Cisco and Juniper have a combined 85,000 employees and $154 billion in market value. The same types of back-end systems will exist in the robotics industry. And the big five countries will benefit from being home to the high-paying jobs and wealth accumulation that go with being out ahead of the 191 other countries around the world. They will produce the Ciscos and Junipers of robotics.

Interestingly, less developed countries might be able to leapfrog technologies as they enter the robot landscape. Countries in Africa and Central Asia have been able to go straight to cell phones without building landline telephones, and in the same way they might be able to jump ahead in robotics without having to establish an advanced industrial base.

The African Robotics Network (AFRON) offers a good model. A community of individuals and institutions, AFRON hosts events and projects to boost robotics-related education, research, and industry on the continent. Through initiatives like its 10 Dollar Robot Challenge, AFRON encourages the development of extremely low-cost robotics education. One winner was RoboArm, a project from Obafemi Awolowo University in Nigeria whose armlike structure is made out of plastic and runs on scavenged motors. The ability to generate low-cost innovation based on scarcity of materials is rooted in the concept of frugal innovation, which will be discussed in chapter 6.

As robotics starts to spread, the degree to which countries can succeed in the robot era will depend in part on culture—on how readily people accept robots into their lives. Western and Eastern cultures are highly differentiated in how they view robots. Not only does Japan have an economic need and the technological know-how for robots, but it also has a cultural predisposition. The ancient Shinto religion, practiced by 80 percent of Japanese, includes a belief in animism, which holds that both objects and human beings have spirits. As a result, Japanese culture tends to be more accepting of robot companions as actual companions than is Western culture, which views robots as soulless machines. In a culture where the inanimate can be considered to be just as alive as the animate, robots can be seen as members of society rather than as mere tools or as threats.

In contrast, fears of robotics are deeply seated in Western culture.

The threat of humanity creating things we cannot control pervades Western literature, leaving a long history of cautionary tales. Prometheus was condemned to an eternity of punishment for giving fire to humans. When Icarus flew too high, the sun melted his ingenious waxed wings and he fell to his death. In Mary Shelley's *Frankenstein*, Dr. Frankenstein's grotesque creation wreaks havoc and ultimately leads to its creator's death—and numerous B-movie remakes.

This fear does not pervade Eastern culture to the same extent. The cultural dynamic in Japan is representative of the culture through much of East Asia, enabling the Asian robotics industry to speed ahead, unencumbered by cultural baggage. Investment in robots reflects a cultural comfort with robots, and, in China, departments of automation are well represented and well respected in the academy. There are more than 100 automation departments in Chinese universities, compared with approximately 76 in the United States despite the larger total number of universities in the United States.

In South Korea, teaching robots are seen in a positive light; in Europe, they are viewed negatively. As with eldercare, in Europe robots are seen as machines, whereas in Asia they are viewed as potential companions. In the United States, the question is largely avoided because of an immigration system that facilitates the entry of new, low-cost labor that often ends up in fields that might otherwise turn to service robots. In the other parts of the world, attitudes often split the difference. A recent study in the Middle East showed that people would be open to a humanoid household-cleaning robot but not to robots that perform more intimate and influential roles such as teaching. The combination of cultural, demographic, and technological factors means that we will get our first glimpse of a world full of robots in East Asia.

HUMANIZING ROBOTS

The first wave of labor substitution from automation and robotics came from jobs that were often dangerous, dirty, and dreary and involved little personal interaction, but increasingly, robots are encroaching on jobs in the service sector that require personalized skills. Jobs in the service sector that were largely immune from job loss during the last stage of globalization are now at risk because advances in robotics have accelerated in recent years, due to breakthroughs in the field itself as well as new advancements in information management, computing, and high-end engineering. Tasks once thought the exclusive domain of humans—the types of jobs that require situational awareness, spatial reasoning and dexterity, contextual understanding, and human judgment—are opening up to robots.

Two key developments have dovetailed to make this possible: improvements in modeling belief space and the uplink of robots to the cloud. *Belief space* refers to a mathematical framework that allows us to model a given environment statistically and develop probabilistic outcomes. It is basically the application of algorithms to make sense of new or messy contexts. For robots, modeling belief space opens the way for greater situational awareness. It has led to breakthroughs in areas like grasping, once a difficult robot task. Until recently belief space was far too complex to sufficiently compute, a task made all the more difficult by the limited sets of robot experience available to analyze. But advances in data analytics (described in chapter 5) have combined with exponentially greater sets of experiential robot data to enable programmers to develop robots that can now intelligently interact with their environment.

The recent exponential growth of robot data is due largely to the development of cloud robotics, a term coined by Google researcher James Kuffner in 2010. Linked to the cloud, robots can access vast troves of data and shared experience to enhance the understanding of their own belief space. Before being hooked up to the cloud, robots had access to very limited data—either their own experience or that of a narrow cluster of robots. They were stand-alone pieces of electronics with capabilities that were limited to the hardware and software inside the unit. But by becoming networked devices, constantly connected to the cloud, robots can now incorporate the experiences of every other robot of their kind, "learning" at an accelerating rate. Imagine the kind of quantum leap that human culture would undertake if we were all suddenly given a direct link to the knowledge and experience of everyone else on the planet—if, when we made a decision, we were drawing from not just our own limited experience and expertise but from that of billions of other people. Big data has enabled this quantum leap for the cognitive development of robots.

Another major development in robotics arrives through materials science, which has allowed robots to be constructed of new materials. Robots no longer have to be cased in the aluminum bodies of armor that characterized C-3PO or R2-D2. Today's robots can have bodies made of silicone, or even spider silk, that are eerily natural looking. Highly flexible components—such as air muscles (which distribute power through tubes holding highly concentrated pressurized air), electroactive polymers (which change a robot's size and shape when stimulated by an electric field), and ferrofluids (basically magnetic fluids that facilitate more humanlike movement)—have created robots that you might not even recognize as being artificial, almost like the Arnold Schwarzenegger cyborg in *The Terminator*. An imitation caterpillar robot designed by researchers at Tufts University to perform tasks as varied as finding land mines and diagnosing diseases is even biodegradable—just like us.

Robots are now also being built both bigger and smaller than ever before. Nanorobots, still in the early phases of development, promise a future in which autonomous machines at the scale of 10^{-9} meters (far, far smaller than a grain of sand) can diagnose and treat human diseases at the cellular level. On the other end of the spectrum, the world's largest walking robot is a German-made fire-breathing dragon that stands at 51 feet long, weighs 11 tons, and is filled with more than 20 gallons of stage blood. Apparently the Germans have a festival involving it.

Recent advances will continue. It is not just Japan's government that is devoting ever-increasing resources to robotics. In the United States, President Obama launched the National Robotics Initiative in 2011 to stimulate development of robots for industrial automation, elder assistance, and military applications. Run by the National Science Foundation, the program has awarded more than $100 million in contracts. France has initiated a similar program, pledging $126.9 million to develop its industry and catch up to Germany. Sweden has similarly earmarked millions to give out to individuals and corporations through innovation awards such as Robotdalen ("robot valley"), launched in 2011.

The private sector is also investing at increasingly higher levels. Google purchased Boston Dynamics, a leading robotics design company with Pentagon contracts, for an untold sum in December 2013. It also bought DeepMind, a London-based artificial intelligence company founded by wunderkind Demis Hassabis. As a kid, Hassabis was the second-highest-ranked chess player in the world under the age of 14, and while he was getting his PhD in cognitive neuroscience, he was acknowledged by *Science* magazine for making one of the ten most important science breakthroughs of the year after developing a new biological theory for how imagination and memory work in the brain. At DeepMind, Demis and his colleagues effectively created the computer equivalent of hand-eye coordination, something that had never been accomplished before in robotics. In a demo, Demis showed me how he had taught his computers how to play old Atari 2600 video games in the same way that humans play them, based on looking at a screen and adjusting actions through neural processes responding to an opponent's actions. He'd taught computers how to think in much the way that humans do. Then Google bought DeepMind for half a billion dollars and is applying its expertise in machine learning and systems neuroscience to power the algorithms it is developing as it expands beyond Internet search and further into robotics.

Most corporate research and development in robotics comes from within big companies (like Google, Toyota, and Honda), but venture capital funding in robotics is growing at a steep rate. It more than doubled in just three years, from $160 million in 2011 to $341 million in 2014. In its first year of investment, Grishin Robotics, a $25 million seed investment fund, evaluated more than 600 start-ups before coming to terms with the eight now in its portfolio. Singulariteam, a new Israeli venture capital fund, quickly raised two funds of $100 million each to invest in early-stage robotics and artificial intelligence. The appeal for investors is obvious: the market for consumer robots could hit $390 billion by 2017, and industrial robots should hit $40 billion in 2020.

As the technology continues to improve, there is an ongoing debate about just how radically human life will be transformed by advanced robots and whether robots will ultimately surpass us. One view in the debate is that it's inevitable robots will pass us; another is that they can't possibly compete with us; a third is that man and machine could merge. Within the robotics community, the future of technology is wrapped up in the concept of singularity, the theoretical point in time when artificial intelligence will match or surpass human intelligence. If singularity is achieved, it is unclear what the relationship between robots and humans will become. (In the Terminator series, once singularity is achieved, a self-aware computer system decides to launch a war on humans.) Enthusiasts for the singularity imagine that investments in robotics will do more than strengthen corporate balance sheets; they will radically enhance human well-being, eliminating mundane tasks and replacing diseased or aging parts of our bodies. The technology community is deeply divided about whether singularity is a good thing or a bad thing, with one camp believing it will enhance human experience as another camp, equally large, believes it will unleash a dystopian future where people become subservient to machines.

But will singularity occur?

Those who believe that singularity will be achieved point to several key factors. First, they argue that Moore's law, which holds that the amount of computing power we can fit into a chip will double every two years, shows little sign of slowing down. Moore's law applies to the transistors and technology that control robots as well as those in computers. Add rapid advances in machine learning, data analytics, and cloud robotics, and it's clear that computing is going to keep rapidly improving. Those who argue for the singularity differ on when it will occur. Mathematician Vernor Vinge predicts that it will occur by 2023; futurist Ray Kurzweil says 2045. But the question looming over singularity is whether there's a limit on how far our technology can ultimately go.

Those who argue against the possibility of singularity point to several factors. The software advances necessary to reach singularity demand a detailed scientific understanding of the human brain, but our lack of understanding about the basic neural structure of the brain impedes software development. Moreover, while weak artificial intelligence, whereby robots simply specialize in a specific function, is currently advancing exponentially, strong artificial intelligence, whereby robots demonstrate humanlike cognition and intelligence, is advancing only linearly. While inventions like IBM's Watson (the computer designed by IBM that beat *Jeopardy!* champions Ken Jennings and Brad Rutter) are exciting, scientists need a better understanding of the brain before these advances progress beyond winning a game show. Watson didn't actually "think"; it was basically a very comprehensive search engine querying a large database. As robotics expert and UC Berkeley professor Ken Goldberg explains, "Robots are going to become increasingly human. But the gap between humans and robots will remain—it's so large that it will be with us for the foreseeable future."

It's my view that the current moment in the field of robotics is very much like where the world stood with the Internet 20 years ago. We are at the beginning of something: chapter one, page one. Just as it would have been difficult in the days of dial-up modems to imagine an Internet video service like YouTube streaming over 6 billion hours of video every month, it is difficult for us to imagine today that lifelike robots may walk the streets with us, work in the cubicle next to ours, or take our elderly parents for a walk and then help them with dinner. This is not happening today and it will not happen tomorrow, but it will happen during most of our lifetimes. The level of investment in robotics, combined with advances in big data, network technologies, materials science, and artificial intelligence, are setting the foundation for the 2020s to produce breakthroughs in robotics that bring today's science fiction right into mainstream use.

Innovation in robotics will produce advances in degree—robots doing things faster, safer, or less expensively than humans—and also in kind: they'll be doing things that would be impossible for humans to do, like allowing a sick, homebound 12-year-old to go to school, or giving those who are deaf and mute the power of speech.

STEP ON IT, ROBOTIC JEEVES

People have been thinking about building driverless cars for almost as long as cars have been around. General Motors introduced the modern concept of the driverless car at the 1939 World's Fair in New York, conceiving of a radio-guided car that could be developed alongside a modern highway system. Then in 1958, GM developed the first driverless test car, the Firebird, which would connect to a track wired with electrical cable. When hooked up with other cars, the system would let each know how much distance to give the others—not unlike the famous cable cars of San Francisco, which use a similar method to propel themselves and maintain safe distances.

But prior to the 2000s, the driverless car remained little more than a futuristic concept. As Sebastian Thrun, founder of the Google Car Project, explained, "There was no way, before 2000, to make something interesting. The sensors weren't there, the computers weren't there, and the mapping wasn't there. Radar was a device on a hilltop that cost $200 million dollars. It wasn't something you could buy at RadioShack." His Google colleague Anthony Levandowski described the shortcomings of the earlier electric models in this way: "We don't have the money to fix potholes. Why would we invest in putting wires in the road?"

Today, however, almost every major car company is researching and building its own version of a driverless car. But the company at the forefront is not a traditional car company at all: it's Google. For the past six years, the tech giant's moon shot development lab, Google X, has been working on the driverless Google car. While much of the technology is proprietary and secret, the company has disclosed a few of its most prominent features. Among other technologies, the Google car includes radar, cameras to ensure that cars stay within lanes, and a light detection and ranging system. Infrared, 3D imaging, an advanced GPS system, and wheel sensors are also being incorporated.

But why would Google get into the car-making business in the first place?

It stems from several important motivations for many of those involved. And it turns out that the development of a driverless car is deeply personal. As Sebastian Thrun explained in a TED talk, his best friend was killed in a car accident, spurring his personal crusade to innovate the car accident out of existence: "I decided I'd dedicate my life to saving 1 million people every year."

Google has hired the former deputy director of the National Highway Traffic Safety Administration, Ron Medford, to be its director of safety for self-driving cars. Medford explained that Americans collectively drive approximately 3 trillion miles per year, and more than 30,000 people die in the process. Worldwide, those statistics are enormous; approximately 1.3 million people die every year in car crashes.

Google, of course, also has an interest in allowing consumers to have more time on their hands—quite literally, to have their hands free. The average American spends 18.5 hours a week driving, and Europeans spend about half that. Any time not spent behind the wheel is time you can spend using a Google product.

But will it work?

There is ample reason to think that robodrivers will be safer than we are now. Accidents are caused by the four Ds: distraction, drowsiness, drunkenness, and driver error. The driverless car promises to reduce all of these significantly. Chris Gerdes, a Stanford professor of engineering, cautions that driverless cars won't fully wipe out human error, but rather will shift it from driver to programmer; that's in all likelihood a significant step forward, especially if a human driver and the programmer can work together. A similar process has unfolded over the years with airplanes, which are now largely driven by autopilot with the pilot still stepping in during crucial times. There remain many gaps to be filled before we can unequivocally say that robodrivers are safer than human drivers. At the top of the list is the software development still to be done to enable robodriving in bad weather and to account for unexpected changes in traffic (e.g., when there's a detour or a police officer directs traffic). But on the whole, given how rapidly progress has occurred and how well the Google car has been shown to perform in clear weather, it's likely that at least partial-robotic driving will arrive in the near future.

The feasibility of the Google car depends on a range of technological, legal, safety, and commercial considerations. Will the technology work? Will it actually make the roads safer? Will people trust and purchase it? Will it even be legal?

These are not academic questions. While only California, Florida, and Nevada have passed laws as of 2013 permitting autonomous cars on the roads, these already represent huge driving cultures and markets. The driverless car has the potential to fundamentally disrupt the modern automotive industry and all of its various branches. As with every other development in robotics, many people will gain—some, like Google's executives and shareholders, may gain immensely—but it's inevitable that others will be displaced. Technology companies have already challenged the automotive market. Uber, the mobile app that connects passengers with drivers for hire, has turned the taxi market on its ear. But what happens when that market is challenged by robots? Uber has already built a robotics research lab stuffed with scientists to "kickstart autonomous taxi fleet development" so they can go driverless. At last count, there were 162,037 active drivers in the Uber fleet who would be kickstarted into obsolescence.

In the United States and many other countries, taxi drivers are often immigrants or others hustling their way up the socioeconomic ladder. It's also a job with tremendous amounts of human interaction. Cab drivers are a great source for every new diplomat or lazy journalist. Conversations with a taxi driver can help assess the national mood, determine what the politics are, or just find out what the weather will be. I suppose a robot can tell you all this—probably with more precision. But will we lose the human touch? More to the point, even if passengers end up preferring robot drivers to humans, what happens to the human taxi driver who loses his job because service industry jobs are at risk in the next wave of innovation as never before?

This isn't just about taxi drivers; the delivery driver may be replaced by Amazon's airborne delivery drones or automated delivery trucks. UPS and Google are also testing their own versions of the delivery drone. Two and a half million people in the United States make their living from driving trucks, taxis, or buses, and all of them are vulnerable to displacement by self-driving cars. It's hard to wrap your head around all the changes this might mean. I met the CEO of a company that develops high-tech access control systems (like the new parking garage system at the airport that tells you how many open spaces are available on each floor) and asked him what worries him about the future. He cited a disruption that I'd never considered before: what driverless cars might mean for parking garages. Would the cars just drive themselves back home and come back when needed? Why have your car sit in a garage and have to pay for it?

The degree to which delivery drones fill the sky or driverless cars fill the streets will eventually be determined not by whether it is feasible technologically and economically—at some point it will be—but by whether humans accept the changes they bring about. Who would you rather trust behind the wheel: a friend, a parent, a person—or a black box that you can't control? Even though accidents happen every day with cars, would we be willing to accept the same from a software glitch? Judging by how much scrutiny each plane crash receives, probably not. If there were a pile-up on the highway because of a software glitch, there would be calls to take the system offline. Such a thing happens every day with human drivers. We have grown to accept that driving leads to more than 1 million deaths a year. Would we accept a computer-based system that produces tens of thousands or hundreds of thousands instead from driverless cars? Probably not. The driverless system will have to prove to be nearly perfect before it scales.

THE MACHINE OF ME

Robots are also beginning to play an important role in the operating room, another place with zero tolerance for error given the life-and-death stakes. In 2013, 1,300 surgical robots were sold for an average cost of $1.5 million each, accounting for 6 percent of professional service robots and 41 percent of the total sales value of industry robots. The number of robotic procedures is increasing by about 30 percent a year, and more than 1 million Americans have already undergone robotic surgery.

The medical applications of robotics are varied. There's the da Vinci surgical system manufactured by Intuitive Surgical in the United States. It's a minimally invasive remote robotic system created to assist with complex surgeries such as cardiac valve repair and is used in more than 200,000 surgeries a year. The robot translates a surgeon's hand into more precise "micromovements" of the robot's tiny instruments. But at a cost of $1.8 million, it's only available to the wealthiest hospitals and institutions. Then there is the Raven, designed for the US Army, a newer surgical robot that can test out experimental procedures. At $250,000, it's a much more accessible option than the da Vinci system, and it's the first surgical robot to use open-source software, which could allow for lower-cost telesurgery systems.

Johnson & Johnson's SEDASYS system automates the sedation of patients undergoing colonoscopies, easing the over $1 billion cost of sedation each year. The services of anesthesiologists typically increase the price of surgery by $600 to $2,000. SEDASYS, already approved by the Food and Drug Administration and going into hospitals today, would cost only $150 per procedure. It would not eliminate anesthesiologists altogether. Instead, like autopilot, systems like SEDASYS merely aid the doctor, enabling an anesthesiologist to monitor ten procedures taking place simultaneously as opposed to having an anesthesiologist in each operating room.

Beyond aiding in existing procedures, robots will even be able to reach places that human surgeons cannot. Ken Goldberg's research team is working on treating cancer with robots that could be temporarily inserted into the human body to release radiation. Instead of radiation from an external source, which damages healthy living tissues along with cancer, these robots release a radio beam inside the body that emits radiation into cancer cells with pinpoint accuracy. Using 3D printing, a medical engineer can even create a customized implant that can travel through a patient's body to fit perfectly where it's needed.

Despite the promise of robot-assisted surgery, it is important not to jump to techno-utopianism. Allegations of unreported injuries from robotic surgery are troublingly common. The *Journal for Healthcare Quality* has reported 174 injuries and 71 deaths related to da Vinci surgeries. With the pressure on insurance companies and health care providers to lower costs, I worry that there will be market forces pushing robots into the operating room at times when a patient is better served by a human being. Robots can eventually improve outcomes in health care, but it would be a human failing if we rush to Doctor Robot due to financial considerations alone.

Robots are also having an impact in the medical field outside the operating room. Across the globe, 70 million people have severe hearing and speech impairments. There is rarely a medical solution to being deaf or mute, and people with these disabilities often live at high levels of social exclusion. While I was traveling in Ukraine, a group of engineering students in their twenties showed me a shiny black-and-blue robot glove called Enable Talk that uses flex sensors in the fingers to recognize sign language and translate it to text on a smartphone via Bluetooth. This text is in turn converted to speech, allowing the deaf and mute person to be able to now "speak" and be heard in real time. With advances like Enable Talk's robot inserts and robot sensory enhancement, robotics might not just aid medicine; the distinction between human and machine itself could start to become blurred.

We can see this line start to blur at Greenleaf Elementary School in Splendora, Texas, where a 12-year-old boy named Christian was diagnosed with acute lymphoblastic leukemia. Because his immune system was compromised, he could not attend school. Instead, a VGo robot, made by a company in New Hampshire, sits in the front row of class for him. The robot has a network-enabled video camera, allowing Christian to sit in his living room and from his laptop see and hear what is happening in class in real time. He can raise his hand (which VGo does for him), be called on by the teacher, and answer a question that the teacher and whole class can hear through the speakers on the robot. Through his robot, Christian leaves the building for fire drills. He walks the halls and stands in line with the students. And students talk to Christian, the sick, homebound 12-year-old, by talking to his robot.

A French robotics company, Aldebaran, has created another interesting use for robots in the classroom: a less than two-foot-tall humanoid robot called NAO that is serving as a teaching assistant in science and computer science classes in 70 countries. It has also been adapted to serve as a classroom buddy to help students with autism communicate more effectively. At an elementary school in Harlem, the NAO robot sits or stands on students' desks and helps them with their math work, all while a professor from Columbia University's Teachers College (who got her PhD from Keio University in Japan) monitors and studies the interactions and pedagogy.

Ten years ago, the advances now entering operating rooms and classrooms would have been nearly impossible to foresee. As researchers, entrepreneurs, and investors think about new applications of robotics, they are no longer considering only tasks that could be done more efficiently by a machine than a human. They are thinking more and more about doing things that humans could never have imagined doing on their own—like Ken Goldberg's radiation-emitting nanobots or Honda's Walk Assist robot that enables otherwise wheelchair-bound people to walk.

Another idiosyncratic but vivid example can be seen in South Korea, where fishermen had long been powerless to deal with the negative impact of jellyfish on their businesses. Jellyfish cost the world's fishing and other maritime industries billions of dollars annually—$300 million in South Korea alone. Then the Urban Robotics Lab at the Korea Advanced Institute of Science and Technology created JEROS—the Jellyfish Elimination Robotic Swarm—a large, autonomous blender that hunts and kills jellyfish at a rate of up to one ton of jellyfish every hour.

ROBOTS AND JOBS

While robots are doing certain things that humans could never do, their main use continues to be work that humans have been doing occupationally for centuries. The term *robot* was coined in a 1920 play, *Rossum's Universal Robots*, by the Czech science-fiction writer Karel Čapek. But its name betrays deeper historical roots. *Robot* derives its etymological roots from two Czech words, *rabota* ("obligatory work") and *robotnik* ("serf"), to describe, in Čapek's conception, a new class of "artificial people" that would be created to serve humans.

Robots in essence represent the merger of two long-standing trends: the advancement of technology to do our work and the use of a servant class that can provide cheap labor for higher classes of society. In this light, robots are a sign of technological advancement but also an updated version of the slave labor that in past centuries people used to exploit other human beings.

The next generation of robots will be mass-produced at declining costs that will make them increasingly competitive with even the lowest-wage workers. They will dramatically affect employment patterns as well as broader economic, political, and social trends. An example can be seen with Foxconn, the Taiwanese company that manufactures your iPhone, along with many other gadgets developed by companies like Apple, Microsoft, and Samsung. Its largest factory complex, in the Shenzhen manufacturing zone near Hong Kong, employs half a million workers in 15 separate factories.

Perhaps thinking ahead about both the economics and the sociology of his business, Foxconn's founder and chairman, Terry Gou, announced a plan in 2011 to purchase 1 million robots over the next three years to supplement the approximately 1 million human workers he employs. Gou has come under fire for his factories' poor working conditions and labor mistreatment. Many workers live inside the factory itself and work up to twelve hours a day, six days a week. But what happens to Gou's 1 million human workers when they have 1 million robot coworkers? While the robots are designed to work alongside humans, they're also designed to keep Gou from having to hire more humans, effectively ending job creation in his factories.

Right now, Gou's robots are slated to take over routine jobs like painting, welding, and basic assembly. Each of these robots currently costs $25,000, about three times a worker's average annual salary, although the Taiwanese firm Delta plans to sell a similar version for $10,000. By the end of 2011, Foxconn had 10,000 robots, or one for every 120 workers, in its facilities. By the end of 2012, the number of robots had jumped to 300,000, or one for every four workers. Gou hopes to have the first fully automated plant in operation in the next five to ten years.

Why would Foxconn make such a massive investment in robotics? Some of it may have to do with Gou's peculiar management style. As he explained in a 2012 *New York Times* article, "As human beings are also animals, to manage one million animals gives me a headache." But Gou is also responding to pure market forces as well. For the past ten years, Gou was able to amass such a large workforce because labor in China has been so cheap. But wages in China have risen along with its overall economic growth—wages for manufacturing jobs have soared between fivefold and ninefold in the past decade—making it increasingly expensive to maintain a large Chinese labor force.

Boiled down to economic terms, the choice between employing humans versus buying and operating robots involves a trade-off in terms of expenditures. Human labor involves very little "capex," or capital expenditures—up-front payments for things like buildings, machinery, and equipment—but high "opex," or operational expenditures, the day-to-day costs such as salary and employee benefits. Robots come with a diametrically opposed cost structure: their up-front capital costs are high, but their operating costs are minor—robots don't get a salary. As the capex of robots continues to go down, the opex of humans becomes comparatively more expensive and therefore less attractive for employers.

As the technology continues to advance, robots will kill many jobs. They will also create and preserve others, and they will also create immense value—although as we have seen time and again, this value won't be shared evenly. Overall, robots can be a boon, freeing up humans to do more productive things—but only so long as humans create the systems to adapt their workforces, economies, and societies to the inevitable disruption. The dangers to societies that don't handle these transitions right are clear.

I anticipate that the same kind of protest and labor movements that advocated against free trade agreements in the 1990s will form in the 2020s once robots begin to really make their presence known in the workplace. The degree to which these robots look more lifelike because of advances in materials science will only make the response angrier and more scared. I got a glimpse of this during violent protests in my adopted hometown of Baltimore in spring 2015. The national and international media portrayed the protests as being about race-based police brutality. Those of us in Baltimore knew it was about more, though. While the triggering event was the death of a 25-year-old African American man in police custody, the protesters themselves consistently rooted their cause and rallying cry of "Black Lives Matter" in more than police brutality. It was about the hopelessness that came from growing up poor and black in a community that had been laid to waste with the loss of Baltimore's industrial and manufacturing base and then gone ignored. Black working-class families had effectively been globalized and automated out of jobs. Many barely hold on with low-paying service industry jobs.

In industrialized countries, what we have witnessed in terms of manufacturing job loss is repeating itself across the economy. Now service industry jobs are also at risk—precisely the jobs that were shielded from job loss in the last wave of mechanization. During the recent recession, one in twelve people working in sales in the United States was laid off. Two Oxford University professors who studied more than 700 detailed occupational types have published a study making the case that over half of US jobs could be at risk of computerization in the next two decades. Forty-seven percent of American jobs are at high risk for robot takeover, and another 19 percent face a medium level of risk. Those with jobs that are hard to automate— lawyers, for example—may be safe for now, but those with more easily automated white-collar jobs, like paralegals, are at high risk. In the greatest peril are the 60 percent of the US workforce whose main job function is to aggregate and apply information.

When I was growing up, my mom worked as a paralegal at the Putnam County Courthouse in Winfield, West Virginia. Her job largely consisted of rummaging through enormous 15-pound books looking for specific information on old court cases and real estate closings. The books were so heavy and the stacks so high that my mom used to conscript me and my little brother to help her. Even as an unemployed high school student in the pre-Internet world when few people owned a home computer, I remember thinking that a computer should be able to do this job more efficiently. But my mom said, "If that ever happens, I won't have a job." Today my mom's job is largely computerized. I now think the same thing about my dad, an attorney who's still working at age 77 with a storefront legal practice just off Main Street in Hurricane, West Virginia. In the next wave of globalization, his job would be at risk as computers develop the ability to work through the more formulaic aspects of legal practice. The role of the lawyer litigating a case in front of judge and jury is not going to be mechanized. But the majority of what most lawyers actually do—developing and reviewing contracts, preparing stacks of paper in legal language to codify the sale of a house or car—these functions will disappear for all but the largest and most complex transactions.

These are just the tip of the iceberg. Think of those taxi drivers who could be replaced by driverless cars. Panasonic created a 24-fingered hairwashing robot that has been tested in Japanese salons. The robot will likely be installed in hospitals and homes as well. It measures the shape and size of the customer's head and then rinses, shampoos, conditions, and dries the customer's hair using its self-advertised "advanced scalp care" abilities.

Then there are waiters and waitresses. Working as a waiter has been an integral part of the career profile for millions of people around the world. By way of illustration, 50 percent of American adults have spent time working in a restaurant; 25 percent say it was their first job. More than 2.3 million people are currently employed as waiters or waitresses in the United States. There is potential for robots to replace many of these waitstaff jobs over time. It's already happening in trial forms in many restaurants around the world. In Asia, many countries are starting to experiment with adopting robots in their restaurants. The Hajime restaurant in Bangkok solely uses robot waiters to take orders, serve customers, and bus tables. Similar restaurants are cropping up in Japan, South Korea, and China. These robots, designed by the Japanese company Motoman, are programmed to recognize an empty plate and can even express emotion and dance to entertain customers. It's unclear exactly how you tip for good service.

The potential loss of restaurant jobs could mean a lot more than the loss of a paycheck; it could mean the loss of social mobility. Waiting tables is a job often held by those with big dreams but a small bank account. Young people, women, minorities, and those without a college degree disproportionally hold these positions and use them as a leg up in society. Currently youth unemployment in the United States is 12 percent, more than twice the nation's overall average, and it is far higher in most of the rest of the world. If entry-level restaurant jobs are reduced or eliminated, how much harder will it be to get a first job? How about a second?

There are earlier precedents for these types of job declines. MIT professor Erik Brynjolfsson calls it "the great paradox of our era. Productivity is at record levels, innovation has never been faster, and yet at the same time, we have a falling median income and we have fewer jobs. People are falling behind because technology is advancing so fast and our skills and our organizations aren't keeping up." In the previous wave of globalization, bank tellers were largely replaced by ATMs, airline ticket counter workers were replaced by electronic kiosks, and travel agents were replaced by travel websites. The robot era may see an even more extreme blow to the sales sector.

The effect of robots on job loss will be highly differentiated by country. The countries that are best positioned are those that are developing and manufacturing robotics for export, that house the headquarters, the engineers, and the manufacturing facilities. These are nations like South Korea, Japan, and Germany.

Those at the highest risk are countries like China that have relied on cheap labor to build up their manufacturing base. As advances in robotics continue, what has happened to manufacturing jobs in many advanced industrial countries may soon happen to industrializing countries. Even in China, where labor has historically been cheapest, it has become increasingly advantageous economically to start buying robots, as Terry Gou is demonstrating at Foxconn.

How will the Chinese government respond to this development? The Tiananmen uprising was a quarter-century ago, but in the minds of Chinese leaders, it may as well have been yesterday. As China grows, it optimizes first and foremost for stability. Above all else, they don't want political instability rooted in economic hardship. The Chinese don't want Baltimore-style protests.

The Chinese government is taking a two-pronged approach: focusing on developing employment by investing heavily in the industries of the future while keeping labor costs low by continuing a forced urbanization policy. In 1950, 13 percent of China's population lived in cities. Today, roughly half the population has been pushed into cities, and the government aims to push that statistic to 70 percent by 2025. This will mean the forced migration of 250 million people from the countryside to city factories in under a decade. Today China has five metropolitan areas with more than 10 million people and 160 with more than 1 million. By comparison, the United States has two metropolitan areas with more than 10 million people and 48 with more than 1 million. The Chinese government continues its forced urbanization program despite the major environmental, political, and administrative obstacles of doing so, because the goal is to keep the cost of labor low. Absent continued movement of people from rural China to the cities, the cost of labor will continue to go up; it's simple supply and demand. If the cost of labor continues to rise, China will lose its special advantage in the global marketplace. Jobs that previously would have gone there have instead begun to move to even cheaper labor markets like Sri Lanka and Bangladesh.

This solution to the challenge of robotics amounts to little more than a country preparing for the future by doubling down on the past—even when it may no longer suit the current era. It's a strategy that holds little hope for coping with the competitive markets of the future, as can be seen in West Virginia.

West Virginia's economy was rooted in the coal mining industry of the 19th and 20th centuries. Scots-Irish immigrants provided cheap labor, and as the cost of these native Appalachians went up, Italian immigrants and then African Americans were brought in to provide lower-cost labor. But as machines grew cheaper and labor more expensive, employers opted for the machines. After all, machines can't go on strike or get black lung, which killed my great-grandfather, an Italian immigrant who worked in the coal camps. The blue-collar workers who traditionally fueled the economy lost their jobs, and the economy fell apart. The state became older and depopulated. The day I was born in 1971, West Virginia had 2.1 million people. Today it has 1.7 million.

The decline of West Virginia was, in essence, a failure to convert from an economy rooted in the strength of people's shoulders to one increasingly mechanized and information based. As much coal is being extracted in the hills of West Virginia today as was extracted decades ago, but the number of workers employed in the mines has plummeted. In 1908, 51,777 workers were employed in West Virginia mines; today only 20,076 people work the mines. Foxconn's employees are the coal miners of today's economy.

Robots will produce clear benefits to society. There will be fewer work-related injuries; fewer traffic accidents; safer, less invasive surgical procedures; and myriad new capabilities, from sick, homebound children being able to attend school to giving the power of speech to those who are deaf and mute. It is a net good for the world. The same can be said of globalization more broadly. It has increased wealth and well-being for people all over the world, but the states and societies (like my native West Virginia) that did not redirect their labor force toward growing areas of employment have foundered.

I think back to the men I worked with on the midnight janitors' shift. Forty years ago, they would have had better-paying jobs in the coal mines or factories. By the 2020s, they might not be able to make a living even by pushing a mop. Right now at the Manchester Airport in England, robot janitors use laser scanners and ultrasonic detectors to navigate while cleaning the floors. If the robot encounters a human obstacle, it says in a proper English accent, "Excuse me, I am cleaning," and then navigates around the person.

How societies adapt will play a key role in how competitive and how stable they are. The biggest wins from new technology will go to the societies and firms that don't just double down on the past but that can adapt and direct their citizens toward industries that are growing. Robotics is one of them, and the others are the very focus of this book. That is why China is not just relying on forced urbanization to produce low-cost labor; it is also investing heavily in the industries of the future. There needs to be investment in growing fields like robotics but also a social framework that makes sure those who are losing their jobs are able to stay afloat long enough to pivot to the industries or positions that offer new possibilities. Many countries, particularly those in Northern Europe, are strengthening the social safety net so that displaced workers have hopes of reemerging in a new field. That means taking some of the billions of dollars of wealth that will be produced from the field of robotics and reinvesting it in education and skills development for the displaced taxi drivers and waitresses. The assumption with robots is that they're all capex, no opex, but the capex you spend on robots doesn't get rid of the opex that people still require. We need to revise that assumption to account for the ongoing costs of keeping our people competitive in tomorrow's economy. We aren't as easy to upgrade as software.

TWO

THE FUTURE OF THE HUMAN MACHINE

The last trillion-dollar industry was built on a code of 1s and 0s. The next will be built on our own genetic code.

Lukas Wartman is the kind of guy you invite to a dinner party to impress your guests. He mixes advice about which Diego Rivera murals to see in Mexico City with accounts of the latest developments in cancer research now taking place inside the world's most advanced life sciences labs. Raised 45 minutes outside Chicago, Wartman speaks with midwestern affability. He's quiet and earnest, with a round face, kind blue eyes, and short brown hair. His Facebook page is filled with photos of him and his dog, Kazu. He's a low-key guy. Even while wearing his white lab coat, the 38-year-old Wartman is reluctant to tout his own expertise or share his remarkable life story.

But Wartman's life *is* remarkable. He works on the cutting edge of genomic technology. From his lab at Washington University in St. Louis, the oncologist and medical researcher studies leukemia in mice, creating comprehensive genomic models of the disease. Even more remarkable, Wartman has battled acute lymphoblastic leukemia (ALL) and survived. Three times.

It is a cruel coincidence that Wartman's favorite class in medical school was hematology, where he looked at leukemia slides under the microscope. He loved the work. "I think I would be a leukemia doctor even if I had no personal experience with it," Wartman says. "You could diagnose a patient's cancer just by looking at the blood smear or bone marrow under a microscope. There's something very satisfying about being in this position, being able to diagnose a cancer by directly looking at it rather than by just taking care of patients."

Wartman has been at Washington University for most of his career. He completed college, medical school, and his residency at the St. Louis university.

Washington University also saved his life, against all odds. In children, ALL is treatable, but it is often fatal in adults. Survival rates for a first relapse are slim, and data for double relapses doesn't even exist. So when Wartman developed ALL for a third time in 2011 when he was 33, no known treatment could save him. His colleagues at Washington University's genomics institute knew the odds were against Wartman's surviving, but they wanted to do something—anything—to save their colleague. They decided to do something never done before: sequence both the deoxyribonucleic acid (DNA) and ribonucleic acid (RNA) from Wartman's cancer cells, then sequence DNA from Wartman's skin sample as well, so they could compare the DNA between his healthy cells and leukemia cells.

All cancers begin with damaged DNA. The DNA becomes damaged through time, or inherited genetic makeup, or environmental factors like cigarette smoke—and as a result it mutates. With cancer, the mutated DNA and RNA, which generally work together to make proteins, are malfunctioning. They are failing to control the growth of unhealthy cells (creating a tumor) or failing in their role as the body's repair engine and allowing cells to become cancerous.

To treat someone like Wartman, scientists want to know whether the protein is malfunctioning because the DNA is providing bad genetic programming or if the RNA's role in creating a protein is not working. Sequencing Wartman's healthy genes, cancer genome, and RNA was a way of pinpointing where the breakdown had occurred.

To do this, the Washington University team ran Wartman's samples through the university's 26 sequencing machines and a supercomputer. Sequencing machines can be as small as a desktop computer or as big as a jumbo Xerox copying machine from the 1980s that takes up half the mailroom. The lab put all of them to work, and they ran day in and day out, zeroing in on the invisible contours of one man's genetic makeup. After several weeks, Washington University's sequencing machines found the culprit. It turned out that one of Wartman's normal genes was producing large quantities of FLT3, a protein that was ultimately spurring his cancer's growth.

Genome sequencing can be a vexing endeavor. Even when sequencing can pinpoint the offending genetic mutation, it's often the case that the medical community does not yet have any drugs or treatments that are capable of targeting the problem, especially if the mutation is rare. But in Wartman's case, there was good news. The pharmaceutical giant Pfizer had recently released a drug, Sutent, that could inhibit FLT3. Sutent was intended for treating kidney cancer, but because of his sequencing, Wartman would become the first person to use it for ALL.

Within two weeks of taking the drug, Wartman was in remission. Soon after, he was in good enough shape to receive a bone marrow transplant to ensure that the cancer would not come back in a mutated form. Four years later, Lukas Wartman's cancer has not returned.

He has had side effects from his treatment. He has eye problems and gets mouth infections. But as Wartman makes clear, it's a small price to pay for being alive. His recovery, by all estimations, is remarkable, though he's not out of the woods. His doctor characterizes his prognosis as "guarded," meaning the eventual outcome is unknown and his condition will remain closely monitored. That he has lived as long as he has, Wartman says, he owes to intensive genetic sequencing. "I don't have any doubt about that at all. In my case, sequencing really saved my life."

Lukas Wartman's story is rare, but his treatment is just the beginning of the potential of genomics. Lukas's story will someday be ordinary—someday soon.

GENOMICS: MELTING CANCER AWAY

Over the past half century, we've witnessed unparalleled advances in the life sciences. Artificial hearts, new wonder drugs, organ transplants, and other developments allow people to live longer, healthier lives.

As Lukas Wartman's story hints, these advances may be dwarfed by the innovations yet to come. In the years ahead, we will live in a world where we'll be able to target cancer cells with true precision, breathe air out of lungs transplanted from farm animals, and deliver medical treatment from the best hospitals in the world to the poorest, most remote corners of the earth.

Genomic research has been racing ahead ever since Gregor Mendel, a Czech monk, discovered the foundations of heredity in the mid-19th century. But the breakthrough that launched genomics on a collision course with medicine occurred in 1995, when the genome of a living organism—*Haemophilus influenza*, a bacterium that causes severe infections, typically in children—was sequenced for the first time.

Almost immediately, the holy grail of genomics came into focus: sequencing the entire human genome. If we could unravel the 3 billion base pairs that make up our DNA and decode who we are on a molecular level, one day Lukas Wartman's doctors at Washington University would be able to understand why and how his cancer was growing.

The announcement of the completion of the first "rough draft" of the human genome was made in June 2000 by President Bill Clinton, and three years later the International Human Genome Sequencing Consortium announced that it had finished the job. The cost of mapping that first genome was $2.7 billion. Over the next ten years, the cost dropped "a million-fold," according to Eric Lander, a pioneering researcher of human genomics. Lander helped sequence the human genome and is now the founding director of the Broad Institute, a biomedical and genomic research center structured as a joint project of MIT and Harvard University. He has a curly mess of graying hair framing a friendly face and bushy mustache. He told me that he thinks the price of mapping the human genome will continue to drop at a shockingly fast rate, allowing for a process of commercialization to take place that unleashes private sector investment into the creation of new diagnostics, therapies, and drugs based around genetics.

The size of the genomics market was estimated at a little more than $11 billion in 2013 and is going to grow faster than anyone could imagine. Ronald W. Davis, director of the Stanford Genome Technology Center and professor of biochemistry and genetics at the Stanford School of Medicine, likens the state of genomics today to that of e-commerce in 1994, the year Amazon was founded and before the founders of Google had even begun working, as students, on Internet search. In addition to the falling cost of sequencing, Davis also cites our increasing ability to draw knowledge out of the genome's terabyte of data as a driver for the coming boom.

The person who helped me understand the potential for new products and businesses built around genomics is someone I met on the racquetball court in downtown Baltimore, Bert Vogelstein. A bit disheveled, even by academic standards, Bert wears a knee brace over baggy gray sweatpants and schleps his racquetball gear to the gym in a dingy old Samsonite suitcase. For the longest time I thought he was just a scraggly gym rat in his sixties. It turns out he is a Johns Hopkins oncology and pathology professor and an expert on cancer and genomics. He's also one of the most cited living scientists in the world.

In the past 40 years, he has published more than 450 scientific papers and been cited more than 200,000 times.

In the 1980s, Vogelstein and his colleagues effectively proved how DNA mutations turn into cancer. Since then, as a result of his work, more than 150 genes have been identified as key actors behind the development and spread of cancer. After proving the relationship between damaged DNA and cancer, Vogelstein began an intensive period of investigation into the meaning of this correlation, trying to figure out how to detect cancers earlier and earlier in their development so that they can be treated before becoming incurable.

His latest effort is what he calls a "liquid biopsy." A blood sample is taken and tested for the presence of even the tiniest amounts of tumor DNA. A tumor detected by Vogelstein's liquid biopsy can be detected at just 1 percent the size of what is necessary to be detected by an MRI, currently the most reliable tool for finding cancer. The amount can be so small that the cancer is discovered even before any symptoms have developed. What this effectively means is that getting a blood test for cancer could become part of everybody's annual medical checkup if the price goes down far enough, as Vogelstein believes it will. The testing done to date by researchers at two dozen medical institutions shows that Vogelstein's method found 47 percent of earliest-stage cancers. While there's still room for improvement, even these early steps offer a remarkable advance over current screening methods. Vogelstein says, "If there were a drug that cured half of cancer you'd have a ticker-tape parade in New York City." Vogelstein's goal is a world where cancers are found and treated, the vast majority of the time, before they pose a mortal threat.

Vogelstein's key partner in the development of the liquid biopsy is his Johns Hopkins colleague Luis Diaz, a 45-year-old scientist who looks strikingly like Robert Downey Jr.'s more academic brother. In addition to his work with Vogelstein on liquid biopsy, Diaz has developed a technique for a molecular pap smear that can detect ovarian and endometrial cancers at an earlier stage. When ovarian cancer is caught at stage 1, when it hasn't spread beyond the ovaries, there's a 95 percent cure rate, Diaz says. But by stage 4, when the cancer has spread beyond the ovaries, that cure rate plummets to 5 percent. The problem, Diaz says, is that most cancers are discovered in stages 3 and 4. Better genetic diagnostic testing will allow doctors to catch cancers in their earliest stages and treat them with much higher cure rates, and Vogelstein and Diaz are already well on their way to introducing tests that could save millions of lives.

Initially Vogelstein and Diaz made respectable progress in their efforts, but in order to move to a stage where their work was treating the many people who need it rather than just functioning in the lab for academic purposes, they realized they needed to bring in the power of private-sector market forces. So in 2009, Diaz and some Johns Hopkins colleagues launched Personal Genome Diagnostics, PGDx, for which Vogelstein serves as a "founding scientific advisor." PGDx now offers cancer sequencing similar to what Lukas Wartman underwent, and it also has a research arm.

PGDx's offices sit on the waterfront in East Baltimore. Diaz and a dozen colleagues have been there for some time, yet the offices are surprisingly barren: everyone is too busy splicing tumors and crunching data to decorate the walls. They are on a mission, one that springs to life when Diaz thunders through the office.

"Cancer right now is prime for the value in genome sequencing," Diaz declares from PGDx's large, minimalist conference room. Speaking about the company's origins, he said, "We saw the requests coming to our group wanting to do more sequencing from patients, from VIPs, and it became clear that we couldn't do it on top of [our research]. We're a research lab, so we needed professionals. We saw a need."

If you're diagnosed with cancer, PGDx can become your cancer specialist. Your oncologist sends in your tumor sample and a spit vial, allowing your cancer cells and normal cells to be compared. Once PGDx gets your samples, its scientists work their genomic sequencing magic. They clean your samples, scrubbing them down so they're ready to take a lengthy ride in a PGDx sequencing machine. Once in the machine, the samples silently churn for hours as every last bit of your DNA is crunched into data. And when the sequencing is done, your DNA's output is hundreds of gigabytes of information—big data now—waiting to be analyzed.

Any genomic sequencing company can do this kind of preparation. What sets PGDx apart is its proprietary computer program, developed at Hopkins, which functions as a high-speed detective. It parses out exactly where proteins are mutating. It makes sense of why your cancer is growing. It gives you more information about your tumor than any oncologist can.

When things go right, the PGDx team can tell you why you have cancer and which medicines might stop these mutations. Sometimes a drug is already on the market. Sometimes there's a clinical trial phase under way for a promising drug. But often the right medicines don't exist. "For years, we were studying one gene at a time. And then we could study ten genes at a time, and now we can study 20,000 genes at a time," Diaz says. "For drugs, we develop one drug at a time. So there needs to be some revolution in drug development that will change that, so there will be more drugs than genes." Today there is a complete mismatch between the drug development process and the speed and precision made possible by genomics. It's a problem that Lukas Wartman, knowing how lucky he was that Sutent was already on the market for him, has been devoting his own research efforts toward.

Wartman spends a lot of time thinking about drug development for cancer with the bold aim of doing away with chemotherapy. "Too many people are still dying of cancer," he says. "The success rate of our traditional chemotherapy has not been enough, so I think the key is to understand as much as you can about the disease. And we can do that if it is through a combination of these sequencing technologies. After we've done that, we'll be able to tailor toward individual alterations in cancer cells."

Wartman thinks that individual sequencing does not necessarily mean a unique treatment for every patient "because that would be too much for oncologists to handle," but instead will lead to more specialized forms of treatment. Wartman says that in the future, cancer treatment is going to be completely different. "Traditional chemotherapy will have a very limited role in the treatment of cancer. That's what I hope. We will essentially be using targeted therapies. . . . I also don't think it will take us two decades to get there. I really think that in the next ten years we'll make substantial progress."

Diaz puts it bluntly: "We are going to understand the pathways activated with cancer very well, and we hope that we'll have designer drugs that will be able to *melt the cancer away*. And that's really the goal. . . . That is going to take about twenty to thirty years to figure out, if not faster."

This revolutionary possibility got a big boost in January 2015 when President Obama announced a $215 million investment by the US government in what could eventually be a decade-long, billion-dollar initiative involving a million volunteers to develop "precision medicines" tailored to a specific person's genetics and the characteristics of their tumor. Developing drugs targeted to the genetics of an individual as opposed to just treating every cancer patient with chemotherapy is as unsubtle a change in the practice of medicine as the introduction of anesthesia in the 19th century. It will make today's most cutting-edge treatments look absolutely primitive by comparison.

HACKING THE BRAIN

The field of genomics is expanding well beyond cancer prevention and treatment. More and more researchers and investors are asking, What about our brains? When you bust up your knee, you get knee surgery; you put a Band-Aid on a bloody elbow; soon you'll be able to sequence your cancer like Lukas Wartman was able to. Yet while every other body part is opening up to medical hacking, the human brain is still something of a mystery. The brain is a collection of soft tissue protected from the outside world by a hard skull. But despite this softness, the brain is something scientists are increasingly thinking of in machine-like terms for diagnoses and treatments.

Scientists now want to break the brain's code and begin to leverage genomics to diagnose and treat neurological and mental illnesses.

I have always been half-obsessed with the genetics of psychiatry. I've had too many friends and family struggle with psychiatric illnesses. During my tenure at the State Department, I could see how big a toll mental health issues took on our soldiers and diplomats coming back from assignments in conflict zones. Early in her tenure at the State Department, Secretary Clinton took a major step forward in acknowledging and encouraging mental health care. "Seeking help is a sign of responsibility and it is not a threat to your security clearance," Clinton wrote in an all-staff email encouraging those who needed help to seek it.

The Pentagon followed suit. Defense Secretary Robert Gates announced that troops no longer had to disclose past mental health treatment when applying for security clearances. This had huge implications for the thousands of soldiers returning from Iraq and Afghanistan in need of treatment for mental health issues. They could finally admit to what they were experiencing.

But a problem lingers: the treatments on offer for my friends, family, and all these soldiers and diplomats are largely rooted in yesterday's science and technology.

If you were depressed in the early 1950s or before, your outlook was bleak. You were confined to a mental hospital, often given up on by your family and doctors. Psychotherapy was the most common form of treatment, with occasional electroshock therapy, and efficacy rates were very bad.

Then antidepressants were discovered. These tricyclic drugs reached into the recesses of the brain, treating chemical imbalances. Suddenly there was a medicine that lifted the dark cloud of depression.

For those who took antidepressants and felt better, the world was unlocked. They entered the workforce, got married, and contributed to society in ways that they previously could not.

But these drugs also came with safety and toxicity concerns. Side effects ran the gamut from sedation to death if mixed with the wrong drugs. As the years wore on, antidepressants started to improve, and the side effects lessened. Then came a new generation of antidepressants that transformed the broader world and our understanding of mental health disorders.

Prozac, the first of these selective serotonin reuptake inhibitors (SSRIs), was pitched by pharmaceutical giant Eli Lilly as an easy-to-prescribe "one pill fits all" for those with depression. Following FDA approval in 1987, nearly 2.5 million prescriptions were issued in its first year on the market. The drug largely worked and became Eli Lilly's blockbuster. Fifteen years after it hit the market, 33 million Americans were taking Prozac and other SSRIs that followed: Zoloft in 1991 and Paxil in 1992. By 2008, antidepressants were one of the most common drugs taken by Americans and the most prescribed drugs for Americans under age 60.

Today most medical treatments for depression involve a combination of SSRIs and cognitive therapy, an approach that works to some degree on about two-thirds of depressed patients. But even with the best care in the world, treatment often involves what amounts to educated guesswork. The people I know with depression are regularly adjusting drugs and doses at their doctor's direction. There are a small number of drugs to work with, all variations on a formula that's now over 20 years old, and the doctor prescribes these based on instinct and experience. It's often guess-and-check work, not based on any knowledge about a specific patient's history or how his or her genetics will respond to a specific therapy.

The post-SSRI opportunity for innovation in mental illness is through genomics. My uncle, Ray DePaulo, chairs the psychiatry department at Johns Hopkins. Uncle Ray and the Broad Institute's Eric

Lander are developing a strategy and program to comprehensively map the genes relevant to the field of psychiatry.

The challenge of mental illnesses is that unlike an ailment such as Huntington's disease, which is caused by a single genetic mutation, most mental disorders are caused by many contributing factors. Dozens, possibly hundreds, of genetic risk factors are at play in mental disorders like depression. Because of the layers of the brain, it's not as simple as unearthing a predisposition for cancer or testing a single gene for Huntington's.

Even so, it's a riddle that researchers are beginning to unravel. Lander explains that in recent years "there's been tremendous progress. Just a couple of years ago, the number of genes that we knew about that played a role in schizophrenia was about zero, and now it's about a hundred," and "that's just in the past three to four years." The potential for the work DePaulo and Lander are doing is vast—and mental health patients around the world could improve by leaps and bounds as better drugs are developed.

One interesting opportunity is in the area of suicide prevention. In the United States, 1.4 percent of all people die from suicide, and 4.6 percent of the population has attempted suicide. Uncle Ray's colleagues at Johns Hopkins studied the DNA of 2,700 adults with bipolar disorder, 1,201 of whom had attempted suicide. They identified a gene, ACP1, that produces a protein that appears in excessive quantities in the brains of people who had attempted suicide. The lead researcher, Dr. Virginia Willour, says that "what's promising are the implications of this work for learning more about the biology of suicide and the medications used to treat patients who may be at risk." By digging into the genetics of suicide, it is possible to develop a treatment that reduces the biological impulse to kill oneself. The researchers have identified the culpable gene. What's next is the development of a commercial product that can go to work on the small region in chromosome 2 with the biological pathway where there is too much ACP1. The very idea that someone would take a pill to prevent suicide works against long-held assumptions about mental illness, but this is the future made possible by genomics.

UNINTENDED CONSEQUENCES

There is a dark side to genomics that even the scientists immersed in the field acknowledge. One of the primary concerns, and one of Luis Diaz's own worries, is that as genomics grows more sophisticated, it will begin a process of creating designer babies. "[The genomic] sequence is going to tell people what risks they have. And those risk profiles will tell them, well, you're predisposed to heart disease," says Diaz. "It will tell you that you're going to be five foot four. You're going to weigh probably about 180 pounds. You are going to be one of the top-percent track runners in your class. You will play basketball. You have an aptitude for math."

Diaz goes deeper. "It's also going to unlock things in the brain. So, behaviors: alcoholism, gambling problems, people with a variety of different addictions. It's going to unlock any genetic predispositions. And then it's also going to be able to predict, well, you're going to have curly hair, straight hair, blue eyes, brown eyes. Are you going to lose your hair at an early age? Is your hair going to remain thick throughout your lifetime? Those types of things. All that will be at our fingertips. So that's scary. Let's take it a step further. You'll be able to tell that probably ten weeks after conception. So that obviously has huge implications, right?"

It is possible today to take a blood sample from a pregnant woman and reassemble the genome of the fetus. Fetal DNA tests have been used in the past to screen for Down syndrome. With advances in genomics, all the genetics of the fetus are now accessible and will force societies around the world to grapple with the issue of genetic selection.

When my wife took a genetic test during pregnancy testing for Down syndrome, a second follow-up test was required. My wife was sick with worry and speculation. It dominated our lives for a month until the next round of tests confirmed that our son would be healthy. It's hard to imagine how we would have felt if we were told that the fetus was healthy, but were also told all of the baby-to-be's genetic predispositions, right down to what would likely kill him in 70 years. Armed with this information, I cannot help but think that many people will opt for designer babies; that those babies brought to term will have the genetics that map to a parent's best hopes. I also can't help but think that if we know a child's predispositions and talents from birth, it will affect decisions about the child's upbringing. Do you give up on college before a child's first day of school? Does fear of a future illness keep parents from allowing a child a normal social upbringing?

Bert Vogelstein and Luis Diaz are concerned that as genetic testing becomes more common, our society won't handle the risk information well. For instance, if a test indicates an increased risk for heart failure, that does not mean "that that is a clinically meaningful risk," Vogelstein explains. As these tests become more common, one of the most crucial issues for us to grapple with, he says, "will be to educate people as well as physicians about the meaningfulness or meaninglessness of particular challenges they might find, and to do it in a fashion that doesn't cause great anxiety or anxiety out of proportion to the risk."

Vogelstein and Diaz's concerns were brought to the surface recently by the genomic testing company 23andMe. Founded by Anne Wojcicki at age 32 in 2006, the company provides ancestry-related genetic reports and uninterpreted raw genetic data for its clients. You spit in a tube, send it to 23andMe's lab, and for $99 they send you back your genetic information. It's not a full sequencing of your genome, but a snapshot of the areas of your DNA that researchers know the most about, like genes that indicate a risk for Parkinson's or how a person might react to certain blood thinners.

Wojcicki, 23andMe's CEO, also happens to be Silicon Valley royalty: she married Google cofounder Sergey Brin; her father chaired the

Stanford physics department; and her mother is a journalism teacher at Palo Alto High who rented out the family's garage to graduate students Brin and Larry Page to incubate Google. It was through a 23andMe test that Brin learned he had a genetic mutation that increased his risk of getting Parkinson's to somewhere between 30 and 75 percent, compared to the broader population's risk of 1 percent. Since then, he drinks green tea and exercises a lot, two activities linked with reducing the risk of Parkinson's.

But while it worked for Brin, 23andMe's version of sequencing is a much simpler version of what Lukas Wartman underwent. Wartman had both his cancer and his full genome sequenced. The difference here is important. Whereas the full sequencing of a tumor is intensive and extensive, and even more so to have an entire genome sequenced, 23andMe is neither. It's a much smaller analysis of some genes that have been linked to common diseases.

Wojcicki's 23andMe is just one company offering do-it-yourself genomic tests, but all of them have faced criticism, specifically around their wildly variable genetic feedback. One test might warn of a heightened risk of arthritis and little risk of coronary heart disease, with another test giving the exact opposite results. The difference is one of precision: there is still a vast difference between the quality of information produced by a $99 test and a test that costs several thousand dollars and takes days of processing through supercomputers. This difference has the potential to cause both false worry and false reassurance.

This problem hasn't been lost on the FDA. In late 2013, it demanded that 23andMe stop marketing its product as "health-related genetic tests" because the company didn't have regulatory approval to make those claims. The FDA's public letter to 23andMe said the FDA was "concerned about the public health consequences of inaccurate results."

Since the FDA came down on them, Wojcicki and her company have bowed to the pressure. Now their tests promise only ancestor information and a file with the raw data. An update on their website reads: "We intend to add some genetic-related health reports once we have a comprehensive product offering. At this time we do not know which health reports might be available or when they might be available."

Though Diaz dismisses 23andMe as "a gimmick" due to its limitations, the company has developed a valuable asset in the form of the genetic material from its 900,000 customers, and it has pivoted its business model in a way that may ultimately produce both commercial and scientific victories. Through a partnership with the Michael J. Fox Foundation, 23andMe built what they called the Parkinson's Research Community with genetic material from more than 12,000 Parkinson's patients. This quantity of data is valuable to pharmaceutical companies developing precision medications and led to a $60 million deal for 23andMe with Genentech. As people continue to pay $99 to 23andMe for ancestor information, they will be building a database that 23andMe can commercialize for drugmakers.

Another set of concerns about the rise of medicines rooted in our genetics comes from people who worry that the development of next-generation drugs arising from genomics will reduce people's focus on diet, environment, and lifestyle, which also damage DNA and cause cancer.

There is already a big gap that breaks down along socioeconomic lines regarding diet and lifestyle. I have found that an accurate measure of someone's net worth can be gauged by what he or she orders at breakfast. The first time I had breakfast with a billionaire acquaintance of mine, we met for breakfast at a fancy hotel. I ordered French toast and bacon. The waiter turned to Mr. Billionaire for his order. Blueberries, he said. Just a big bowl of antioxidant-loaded blueberries with a health drink on the side. By contrast, when I was teaching during the 1990s at Booker T. Washington Middle School in West Baltimore, 97 percent of the kids were on the free lunch program and ate snack-sized bags of chips on their walk to school for breakfast.

As we arm ourselves with more and more information about genetic dispositions, I believe that our behaviors are in danger of skewing even further along socioeconomic lines. With information about
his predisposition to Parkinson's, Sergey Brin changed his personal
habits for the better. The sad truth is that in socioeconomically depressed communities like my native West Virginia or the community
where I taught in Baltimore, if people are given data about their genetic predispositions yet don't have the money or medical options
to do anything about it, their health outcomes will feel more predestined. Rather than inspiring behavior change the way it did in Sergey,
genetic testing may cause people who already feel disempowered to
further reconcile themselves to unhealthy lifestyles.

Dr. Ronald W. Davis, director of the Stanford Genome Technology
Center, points to one company—which he advises—that is bridging
genomics and environmental factors.

BaseHealth was started by a former student of Davis's. Its signature product, Genophen, sequences your genome (like many other
companies do) but then uses a "risk engine"—a big data program that
applies algorithms across medical, behavioral, and environmental
information—to serve up a personalized set of behavior and treatment
recommendations to accompany the genetic data about which diseases you have or are predisposed to get. Davis describes Genophen's
work as "rather breathtaking in its ambition."

For me, Genophen would take all my inherited genetic information and combine it with an array of data points from my life, from
growing up in a house filled with cigar smoke to accounting for the
chemicals spewed into the air by Charleston's chemical plants to more
positive factors like the mostly healthy diet my wife insists on and
lots of running and racquetball. These and many, many more details
from my life are put through computer models that mine a mountain
of data inside Genophen's servers and come up with a diagnosis of
what diseases or conditions I have and what I'm likely (or unlikely) to
have in the future. I'd be given a personalized plan of treatments and

behavioral modifications for me to adopt to address the conditions I have and reduce the likelihood of contracting any diseases in the future.

BaseHealth also tries to address Bert Vogelstein's concern, about patients not understanding how to cope with the information they receive, by selling their product only to doctors. The doctors access the Genophen information through a dashboard that only they and their nurses have access to in the doctor's office. The doctor acts as translator and guide, the well-trained human professional helping the patient make the best use of his or her health data.

The costs in the field are out of reach for most people, but they're coming down quickly. A test at Luis Diaz's PGDx costs between $4,000 and $10,000 depending on how detailed the test is. Three years ago, it held a daunting $100,000 price tag.

The ultimate goal, Diaz says, is to bring that cost down even further and to get this kind of testing covered by insurance companies so that it can be adopted on a much wider scale. At the time of his treatment, Lukas Wartman was able to have his cancer sequenced only because his university had a slew of sequencing machines that his colleagues could use free of charge. Had he not worked at a genomics institute and been well liked, he would have died from ALL in his thirties.

Vogelstein is optimistic that this is only a short-term problem. "Certainly in twenty years it will be trivial to sequence the base pairs in a human's DNA," he says. "That can be done now, and the cost is getting close to $1,000, which is the bar that was set several years ago. But in twenty years, it certainly will be considerably less than $1,000. So the technology for doing that is already available, but not scalable to billions of people, but it will be in twenty years. So everyone who wants it, certainly in the developing world, could have his or her genome sequenced."

PIG LUNGS AND WOOLLY MAMMOTHS

There is a specific stripe of entrepreneur who views even the most cutting-edge genomics research taking place in and around academic settings as too unimaginative. These are researchers working on innovations that seem so far-fetched that it is hard to believe they could ever be real.

Enter Craig Venter, a scientist at the National Institutes of Health (NIH) when the Human Genome Project kicked off.

Venter, the second of four kids, grew up in working-class San Francisco. Impatience-in-overdrive has always characterized his personality. He was both motivated and easily bored as a kid, setting records on his high school swim team but almost flunking out with bad grades. And as an NIH scientist working on the Human Genome Project, Venter pieced together a faster way of harvesting information about a person's genome.

When he asked the head of the Human Genome Project for $5 million to sequence the entire human X chromosome using his new technology, he was told that it was too much money. In response, Venter resigned and started a company to compete with the Human Genome Project. He would race them to map the genome, pushing them to work harder, faster, smarter in a race that would eventually be declared a tie.

Venter has recently started two companies that aspire to nothing less than adding decades to our lives. The first company, Synthetic Genomics, announced a project in 2014 that aims to genetically engineer pigs with organs that can be safely transplanted into human beings. The process, called xenotransplantation, involves modifying a pig genome so that a pig embryo can grow up with organs that can be harvested and transplanted into humans. The initial focus is on lungs, but at the time of the announcement, Venter said that hearts, kidneys, and livers could also be possible.

When xenotransplantation works and when it becomes a

mainstream medical intervention, the standard idea of organ donation and transplantation might evaporate. If human-compatible lungs, kidneys, and hearts can be grown in pigs, they won't be a scarce resource any longer. Getting a new kidney might be no more difficult or exotic than knee replacement surgery.

Venter's second genomics company is even more audacious. Human Longevity, Inc. (HLI) aspires to use genetic data to dial back the effects of aging. "Aging Is the Single Biggest Risk Factor for Virtually Every Significant Human Disease" blare HLI's promotional materials. Venter and his colleagues believe that by building the world's largest whole-genome sequencing center, their scientists will be able to develop products that slow the process of aging. Venter's cofounder, Peter Diamandis, says that "between 1910 and 2010 improvements in medicine and sanitation increased the human life span by 50 percent, from 50 to 75 years. Today, with the emergence of exponential technologies such as those being pioneered and advanced by HLI, we have the potential to meaningfully extend the life span even further."

As incredible as Venter and Diamandis's goals may be, they're making rapid and impressive progress. HLI has recently lined up some of the best medical partners in the world, including PGDx. It raised $70 million in venture capital, and after only eight months of existence, it is now, according to one of its investors and board members, the world's largest whole-genome sequencing center.

While Venter is renowned for his almost outlandishly bold ventures into the future of genomics, even he is one-upped by a new branch of research that seeks to defy the impossible—not by merely prolonging life but by bringing the extinct back to life.

A few years back, a team of scientists used the DNA of a dead bucardo, a wild goat indigenous to the Pyrenees Mountains that went extinct in 2000, to create bucardo embryos. These were implanted in 57 regular (as in *not extinct*) goats' wombs. One of the embryos made it to term. That's right: a regular goat gave birth to a once-extinct bucardo goat in 2003. Granted, the bucardo didn't live long; it managed to survive for only several minutes after its birth. But the sheer possibility has nonetheless taken hold in the minds of curious researchers.

In 2012, the Revive & Restore project was started in San Francisco to bring extinct animals back to life using advanced genomics technology. As Revive & Restore sees it, the well-preserved DNA of many extinct animals can bring these animals back to life. As with the bucardo, this means finding the most genetically similar animal and implanting it with the extinct animal's embryo. Efforts are already under way to "deextinct" the carrier pigeon, the health hen, and an Australian frog most notable for giving birth through its mouth.

It's unclear how far this technology could go. A mountain goat is one thing. But what about putting woolly mammoths back on the earth? If it's doable—and even Revive & Restore scientists admit that it probably isn't with current technology—do we want to turn back the clock? This is another case, like the potential to develop designer babies, where advances in science and technology put humans in a godlike role. In the same way in which human behavior has changed the earth's climate, advances in genomics could alter the world's ecology. Species often become extinct for a reason. Reintroducing them will change food chains and could introduce viruses and bacteria that nature has not adapted to contain. As our ability to manipulate life grows stronger, it needs to be governed by our human judgment.

KEEPING UP WITH THE GENOMIC JONESES

There is good reason that most of these otherworldly advances in genomics are coming from the United States, but it is just a matter of time before other countries, notably China among them, begin to catch up.

There are three things necessary to create breakthrough advances in the life sciences: great scientists, lots of capital for academic research, and a venture capital market to help turn academic research into commercial products.

The main reason the United States is ahead today is that leading international scientists are still clamoring to join American universities. Among the world's most highly cited scientists, one in eight were born in developing countries, but 80 percent of them now live in developed countries. The American university system offers foreign scientists many opportunities they don't have back home. A survey of 17,000 scientists in 16 countries found that foreign researchers migrate for two reasons: to improve their career prospects and to join outstanding research teams. The United States does well on both of these tracks, and it has paid off. In every reputable ranking system of worldwide science departments, US science departments dominate. The United States is the number one destination for foreign scientists from nearly every nation.

But for all its dominance in cultivating brainpower and scientific achievement, the United States is not guaranteed to retain that leadership forever. Perhaps no other R&D project has unleashed more scientific knowledge during my lifetime than the Human Genome Project, which brought in scientists from around the world—Australian, British, and French citizens were on the NIH team. These were scientists from familiar, reputable universities. But when Bill Clinton announced the draft human genome, he was also careful to credit a group that had contributed only about 1 percent of the sequencing: the Beijing Genomic Institute.

In the 15 years since the great race to sequence the human genome, China has emerged as a leader in genomic research. No longer just a 1 percent contributor, the Beijing Genomic Institute (since renamed BGI) is now the largest genomic research center in the world, with more sequencing machines than the entire United States. Some of its researchers are in early discussions about eventually sequencing the genomes of almost every child in China.

Since 1998, the share of China's economy devoted to research and development has tripled. While the portion of global research and development (R&D) in the United States fell from 37 percent to

30 percent in the past decade, China's share of global R&D increased from about 2 percent to 14.5 percent. With nearly 2 percent of China's mammoth GDP going toward R&D, it now has the statistical edge over Europe, and the United States is struggling to retain its lead.

While China's investment is climbing, America's R&D investment in recent years has actually been below its 2008 level, a lingering side effect of declining research budgets following the financial crisis. In addition to the foreclosures and unemployment, irresponsible behavior in the financial services sector resulted in less funding for cancer research.

The United States is the second-largest producer of science and engineering academic articles (if you count the 28-country European Union as a single bloc), contributing a quarter of the world's output. But the US numbers have been declining over the past decade. Meanwhile, China's output has been skyrocketing—from 3 to 11 percent— and it is becoming the world's third-largest producer of scientific articles.

R&D funding and scientific articles that become the intellectual basis for research are the building blocks for discovering new cancer-fighting drugs and life-altering medical advances.

China's strategy emanates from the top of the Chinese government. China's State Council has established genomic research as an economic pillar of its 21st-century industrial ambitions. In three years, the Chinese government has successfully attracted 80,000 Western-educated Chinese-national PhDs to return to China.

The Chinese government officials and business leaders I've spoken with on the subject put it in plain strategic terms. They believe that they missed out on the benefits that came with being an early leader on the Internet. One Chinese CEO told me he believes that the wealth and power that came from being the center of the Internet's commercialization extended America's reign as a superpower by ten years. Many of the most powerful Chinese leaders believe that genomics is the next trillion-dollar industry, and they are determined to be its leader. One opportunity being monitored by the Chinese relates to the drug-development process in the United States. If the FDA does not change its drug-development process to speed the delivery of the kinds of personalized medicines made possible by genetic sequencing, as described by Luis Diaz and Lukas Wartman, then patients may go abroad (perhaps to China) for individualized treatment therapies.

At the core of the Chinese strategy are companies and institutes like BGI that live in the gray space between state and private sector. These are nominally private organizations, but they are engorged with capital and blessed with support from China's central authorities, who are determined to see them succeed for the benefit of China. In 2010, BGI got a line of credit of $1.58 billion from the China Development Bank. Today BGI's revenues come from a variety of sources, though they are difficult to verify. One large category of revenue is "anonymous donations." Other sources include providing data analysis to pharmaceutical companies, sequencing genomes for researchers and individuals, and money that is reported from municipal and federal government institutions. BGI also benefits from the low labor costs of its thousands of employees who, on average, earn only about $1,500 a month.

I think the competition coming from China is actually a good thing. The Human Genome Project moved as swiftly as it did because one of its hotshot researchers, Craig Venter, resigned to establish a company to compete with it. I also think that the more capital that floods into basic research in genomics, the faster we'll see results that benefit people all over the world. BGI has been very outward looking relative to Chinese Internet companies, positioning itself as a global company from its earliest days. Some of its collaborations with non-Chinese entities are only for the good, including one with Autism Speaks, a US nonprofit, to sequence the DNA of 10,000 people whose families have autistic relatives.

When it comes to the rest of the world, the story is mixed. As of now, Europe is on the map academically in genetic research but is far from being a peer of China or the United States in commercialization. The top scientists from India, Latin America, and elsewhere tend to end up at US universities or companies even if they remain physically based in their home countries.

The best example of how *not* to position a country for an industry of the future comes from Russia, for reasons that date back to the beginning of the Cold War. Soviet leaders valued a narrow set of scientific research areas. While the Soviet system produced huge numbers of Soviet scientists, their work was driven by the government's political and military priorities. The great space race resulted in major developments in both the United States and the Soviet Union, but at the same time, the Soviet Union's scientific endeavors were hobbled by the regime's ideology.

An example can be seen in Trofim Lysenko, a scientist who climbed the ranks during the Stalin regime. Lysenko railed against genetics as "bourgeois pseudo-science." He believed that organisms' characteristics were shaped by their environment, and these characteristics were passed onto offspring. In Lysenko's scientific view, if you plucked all the leaves off a tree, the next generation of trees would also be leafless, a remarkably Marxist-Leninist approach to science.

Lysenko convinced the Soviet Agriculture Academy that the study of genetics was bad for the country and the cause, so schools removed any references to genetics in their books and curriculum. Soviet scientists learned to fall in line, publishing articles with absurd results that echoed Lysenko's theories. Those who embraced Lysenko-style research landed funding and scientific awards. On the flip side, scientists who disagreed with Lysenko were imprisoned and sometimes executed.

The lack of any meaningful activity in the field of genomics in Russia today dates back to Lysenko. His views were codified under Soviet law in 1948, and Mendelian genetics did not reenter Russian scientific curricula until years after Lysenko's death in the 1970s. The first "ethnically Russian" genome was not sequenced until 2010, using equipment purchased from the United States and BGI.

INNOVATION FOR EVERYBODY

Much of the innovation in the life sciences coming from China, Europe, and the United States is initially benefiting wealthier households and societies. Technologies like genetic sequencing offer exciting new solutions for people like Lukas Wartman, but around the world, millions suffer preventable deaths because they're unable to get access to simple medical information or treatments.

As the Craig Venters of the world race toward cutting-edge breakthroughs, others are seeking to harness the burgeoning telecommunications infrastructure in the developing world to deliver everyday health care needs better. Delivery of medical services will never be equal, but pioneering initiatives to expand access to care across socioeconomic lines are beginning to take hold and improve lives on a huge scale.

The infrastructure that makes this possible is the mobile phone. Six billion of the 7 billion people on earth have mobile phones, more than have access to toilets. During my travels through Africa and low-income areas of Southeast Asia, I have seen mobile phone-based programs that have proven to be effective for a range of health-related interventions, including diagnosis, disease monitoring and compliance, expert assistance for community health care workers, and programs to promote education and awareness. Mobile phones are well suited for these functions because they're nearly ubiquitous, people take their phones with them everywhere, and they're easy to customize with special-purpose applications. Special apps can give access to the phone's hardware (like the camera) and standard applications (like email, calendar, and contact list). And they can allow phones to connect wirelessly to devices like blood pressure monitors, electrocardiographs, and other sensors. A mobile phone cannot sequence someone's genome yet, but it can be used to take a blood sample and transmit the data to a lab on the other side of the world.

One of the most interesting ways mobile telecommunications have been used to address health problems in the developing world comes from a company called Medic Mobile. I got to know 27-year-old Medic Mobile CEO Josh Nesbit while we were in a jungle in Colombia, in the last stronghold of the FARC guerrillas. Josh was a mobile expert we had brought in for a State Department program with the Colombian military, and he was educating local stakeholders about how mobile devices could be used to map land mines in the area and reduce the loss of life and limb.

When Josh was an undergraduate student at Stanford, he worked at St. Gabriel's Hospital in rural Malawi, one of the poorest countries in the world. While working there, he was struck by how far many of the patients had to travel for simple care. A trip to the doctor often meant a 100-mile trek to the hospital. Community health workers frequently walked dozens of miles to turn over handwritten reports. He also noticed that he got better reception on his mobile phone in poor, rural Malawi than he did in California. This is not unusual. The World Health Organization estimates a shortage of 4.3 million health care workers in 57 countries in the developing world, 36 of them in Africa. Meanwhile, mobile telecommunications have covered most of the continent. Josh made the connection and decided that leveraging this mobile infrastructure would be the founding goal of Medic Mobile.

Josh returned to St. Gabriel's and set up a program equipping the hospital's 75 community health workers with mobile phones and training on how to use the phones to allow patients to respond to medical questions and monitor how well patients were adhering to their prescribed course of care. The pilot program ended up saving over 2,000 hours of worker time and doubled the capacity of the hospital's tuberculosis treatment program.

Today Josh and Medic Mobile's big project is to try to develop a tool that uses the light and camera on a mobile phone to diagnose malaria and tuberculosis, and for under $15. In ten years, Josh tells me, "there will be many new types of health workers, all supported by mobile technologies. Health systems will be decentralized, local, and preventative." He adds, "Health equity will exist in many more places, and the biggest gains will come from delivering basic services to the hardest-to-reach communities. Primary health care information will be free for every family, zero-rated by governments and mobile operators. There will be vaccines for malaria, cholera, and other deadly diseases, and mobile technologies will play a critical role in distributing them to everyone."

Dozens of mobile health programs like Josh's have been going through pilot programs over the last five years, since mobile networks began to cover the African continent and usage shot up. A Kenyan company, Shimba Technologies, developed a mobile medical directory and "knowledge app," MedAfrica, to address the health challenges in its home country, where there are only 7,000 registered doctors in a country of 40 million people. The World Bank puts the doctor per 1,000 person ratio of Kenya at 0.2; in the United States, that number is more than ten times larger—2.4 as of 2010.

To address the doctor deficit, Shimba too decided to take advantage of mobile, since 93 percent of Kenyans are mobile phone users. The app has a symptom checker, first aid information, doctor directories, a hospital locator, and alert systems. In a country with vast rural areas and tenuous access to care, MedAfrica offers a new way for any Kenyan with a mobile phone to get some form of health care, pairing medical expertise with villagers all over the country.

In a more specialized effort, a graduate student at MIT started EyeNetra, a company to help bring optometry to the more than 2 billion people in the developing world who cannot access eye tests. EyeNetra has figured out how to mount a plastic lens, called a viewer, on a smartphone. The patient looks through the viewer, which connects to an app that can diagnose nearsightedness, farsightedness, and astigmatism and prescribe corrective lenses as needed. This saves a trip to the doctor and the use of a $45,000 autorefractor machine. Shortly after launching, it successfully conducted more than 30,000 exams and raised $7 million in venture capital from legendary venture capitalist Vinod Khosla, allowing it to expand its reach even further.

This is the future of health care in rural areas. For just about any part of the body or any illness, there is an entrepreneur thinking about how mobile technology might be applied to narrow the divide in access to health services. The life-sciences companies that will succeed over the next 20 years will harness this spreading mobile technology and bring better and more extensive care to every corner of the earth.

A highly connected world will also give rise to new possibilities for medical specialization and globalize the supply chain for medical diagnoses. Dr. Deborah Schrag, chief of the Division of Population Sciences and Medical Oncology at the Dana-Farber/Harvard Cancer Center, believes that real savings and efficiency may be discovered by tapping into the talents of the villagers in places like Bangladesh, many of whom were unable to complete their secondary schooling, much less a four-year medical degree.

With 78 organs, 206 bones, and 640 muscles, not to mention up to 25,000 genes, our bodies are complicated machines. To develop a comprehensive understanding of the body, medical school is necessary. While we have traditionally trained our doctors to be experts on the entire body, Schrag asks a simple question: Can we divide that expertise and train an entire village halfway around the world to become the globe's expert on a single part of the body for one specific disease?

Schrag suggests we look at breast cancer by way of illustration. Women undergo detection via mammograms, and doctors make a diagnosis. But these mammograms are in fact easy to interpret, and it's possible that lower-cost workers somewhere in the world could be trained to do so. They could look at dozens of breast scans and learn how to categorize them according to the Breast Imaging Reporting and Database System from negative to known biopsy-proven malignancy.

Or they could learn how to recognize when something is abnormal, flag it for further investigation, and hand the case over to the white coats.

Aggregating information and creating a treatment plan will still demand the expertise developed through medical school, residency, and internship. Specialists will continue to exist, but their time could be more efficiently spent. They will focus on the special cases that require treatment rather than shuffling through piles of routine cases. Of course, there's also the possibility that with the coming automation boom, this same identification work will be taken up by robots rather than humans.

So what does this mean for the woman who comes in for her annual mammogram? It means faster and cheaper service. She still has to make an appointment, fill out insurance and medical background paperwork, get her exam, and wait for her results to be interpreted. But the wait time will be shorter and the bill will be less expensive. Most health insurance companies pay around $170 for a mammogram, and uninsured women pay an average of $102. By having low-cost subject experts filter normal scans, we could see these costs decline.

This is another dimension of innovation: while the wealthy generally benefit most over the short term, innovations have the potential to become cheaper over time and spread throughout the greater population. If the cost of a mammogram can be vastly reduced, the exams ought to become available to more women. At the very least, that's the hope.

While ideas like these could save health care providers and patients money over the long term, they are hard to implement. Not only is it difficult and expensive to identify and train an unconventional workforce; providers and patients must then be comfortable with and confident in the procedure. Imagine that you or your wife or mother finds a lump in her breast one morning. You rush to the doctor to have things checked out. Fifteen minutes after the exam, the nurse practitioner comes to tell you that the breast cancer experts—not doctors, mind you—said she is all clear. You are free to go. Many people, rightly or wrongly, will want a second opinion. I know I would.

Sustained credibility is the tough sell that hospitals would have to make to patients and family members everywhere until the practice of overseas scan filtering could be normalized. There are not only first-mover advantages. There are distinct first-mover disadvantages. However, American society has accepted similar changes in the past. Think about the rise of nurse practitioners doing medical tasks, like giving vaccines, that in the past were performed only by physicians.

The adoption of new technology finally occurs when ease of use, economic savings, and trust all come together to work toward change.

EVERYTHING WE KNOW ABOUT THE LIFE SCIENCES IS GOING TO CHANGE

Dr. Davis was right. Where we are today with genomics is the equivalent of where we were in 1994 at the advent of the commercial Internet. Genomics is going to have a bigger impact on our health than any single innovation of the 20th century. We will live longer lives, but our lives will grow more complicated as we manage more information and more choices. We will know more about the biology of who we are and what we will be than we can even imagine today.

Accompanying and enabling the development of genomics are connection technologies—those that link us to information and to each other—that are growing more powerful and less expensive. The first beneficiaries of the innovations will be blueberry-eating billionaires, but the process for mainstreaming these advances within and across societies will likely only take 20 years, well within the life span of most of the people reading this book.

The sheer amount of information that lies around the corner makes me think of Mark Twain's words from *Life on the Mississippi*: "There is something fascinating about science. One gets such wholesale returns of conjecture out of such a trifling investment of fact." The life sciences have been a field where we have lived with a trifling amount of fact relative to what will be available in just a few short years. We are going to make Twain's observation the stuff of history and learn more about ourselves in the next 20 years than the sum of what we learned during the centuries prior. Genomics will become a trillion-dollar industry, extending lives and nearly eliminating diseases that kill hundreds of thousands of people a year today.

THREE

THE CODE-IFICATION OF MONEY, MARKETS, AND TRUST

Is there an algorithm for trust? New ways to exchange are forcing a rewrite of the compact between corporation, citizen, and government.

When I was growing up, money was something you put in a wallet. Buying something meant going to a store, talking to a cashier, taking out your wallet, and handing over your bills. I still remember my father's worn, brown leather wallet. Even as a child, I understood its importance. That wallet was how my father would pay for dinner—or for a treat for me. Its size determined the course of each week and each day. When we left for vacation, it would be thick with possibilities. When it grew thin, it was time to head home.

Money has long been primarily a physical entity—something that can be held and weighed. Many of the world's currencies, in their very names, reflect the notion of money as something tangible. The "peso," the Israeli "shekel," and the British "pound" all derive from words for weight. "Ruble" comes from the Old Russian *rubiti*—"to chop, cut, hew"—because the original metallic currency came in silver bars, from which the needed amount was cut off. For a more vivid example, there's a scene in *Goodfellas* where the gangster Henry Hill's wife, played by Lorraine Bracco, asks him for money to shop. He asks her how much she wants. She holds up her index finger and thumb a few inches apart and says, "That much," and Henry hands over a stack of cash.

But in the past half century, the modern financial system has designed a series of conveniences that have allowed us to move away from physical cash. Like many other adults, when I went to college, I got my first credit card. When I began to travel, I took traveler's checks along with me. ATMs—invented in the late 1960s but not popularized until the 1980s—allowed us to bypass the bank teller to access our cash. All that was required was an ATM card and a pin number. Since then, as with robotics and the life sciences, change has come quickly. Online banking began to take off during the mid- to late 1990s. The consumer-to-consumer marketplace eBay was launched in 1995. The online payments service PayPal was established in 1999. Today digital banking has become nearly universal in developed economies, as has mobile banking, with the ubiquity of cell phones. More than half of American adults use mobile banking, and globally, well over half a billion people do. By 2017, that number will jump to 1 billion. For these customers, the mobile phone is eliminating the need for ATMs just as ATMs eliminated the need for bank tellers. The phone is now the bank, and I get downright irritated when I have to dig through my dresser drawer to find my checkbook to write out a check, 20th-century style.

When I was growing up, wealthy people had stacks of cash and thick wallets. Today most wealthy people—and increasingly more of the rest of us—are virtually walletless, or they use a virtual wallet. And while our genomes are being decoded over the next 20 years, our money will be coded—broken down into 1s and 0s and wrapped within powerful tools for encryption. We're still only beginning to discover the possibilities that digital currency will open up. But the code-ification of money, markets, payments, and trust is the next big inflection point in the history of financial services. Understanding what it means for you and your business will be important regardless of whether you are a plumber or the CEO of a Fortune 500 company.

CODED MONEY, SQUARED

To understand the implications of coded money, I went to talk to Jack Dorsey. At 39, Jack is the quintessential entrepreneur. He cofounded Twitter, a company that has revolutionized communication. And while still chairing Twitter, he cofounded a second company, Square, which aims to revolutionize the way we use our money. Jack, CEO of Square and CEO of Twitter as of this writing, is a true visionary. He's deeply attuned to the latest trends and most ambitious ideas of the tech world, yet is still a sweet, soft-spoken Missourian at heart. We met at Square's new office in Mid-Market San Francisco, and Jack quickly broke into the thinking behind Square and his goals for the future of commerce.

With Square, Jack's original insight was to invent a way to make everyday payments using a device that is growing even more precious than our wallets: our mobile phones. Every morning before I walk out the door, I pat myself down to make sure that I haven't forgotten the three most important things I need to make it through the day. My wallet goes in my back left pocket. Keys go in the front right. The front left is reserved for my phone. Jack wants to make it so you can leave the wallet at home.

Square allows customers to pay (and merchants to vend) through their phones or tablet computers. The first iteration involved a little white square device that you could insert into your phone to process a credit card payment—just swipe the card through a reader on the square. New iterations of Square don't even require its namesake square insert. All you have to do is hold out your phone for a cashier to scan; no need to open your wallet. Many brand companies eager to reduce customer wait times have adopted this technology. Go into any Starbucks and witness coffee drinker after coffee drinker using Square to get in, out, and recaffeinated more quickly.

The inspiration for Square came out of a failed transaction. Jim McKelvey, Jack's partner in founding Square, is an entrepreneur and a glassmaker. Among other projects—writing computer programming textbooks, working at IBM, starting a digital publishing company—McKelvey founded a boutique glassblowing company. The company made world-class glass faucets, but that wasn't enough to prevent Jim from losing a $2,000 sale because he did not have the technology to accept an American Express credit card.

In early 2009, Jim shared the story with his old friend Jack just when Jack was leaving Twitter full time and looking for new opportunities. Both were surprised to see just how far credit card technology lagged behind the mobile breakthroughs of recent years, and they began discussing an alternative payments strategy. In December 2009, Jim and Jack launched Square. They released their first product less than a year later and hit their one billionth payment in November 2014. Square is part of a fierce competition for the future of payments, competing with platform companies like Google Wallet and Apple Pay and other start-ups, like Stripe, a San Francisco–based company started by two Irish brothers, which is doing approximately $1.5 billion in transactions annually.

From its inception, Square has been about enabling the kind of small-scale transactions that Jim lost out on in his failed glass sale. Its approach has been to try to eliminate the cost and complication of standard credit card transactions. Typically when merchants accept a credit card, they are assessed two types of fees. First, they pay a series of fees to merchant account services providers, which act as the middlemen between the merchant and the bank that manages the merchant's account. These fees add up: there's a monthly statement fee ($10 a month on average), a monthly minimum fee ($25), a gateway

monthly payment (between $5 and $15), and transaction fees (typically 0.5 to 5 percent per transaction, plus a flat fee of 20 to 30 cents,
which is why many stores set minimum transaction amounts on credit
card purchases). The second category comprises the fees paid directly
to credit card companies. The biggest of these, called the "interchange
fee," accounts for the majority of credit card processing fees.

Sound confusing? It is. Fees vary based on the type of business a
merchant runs, the market power it controls, and the rewards offered
by credit card companies, among other variables. And credit card
companies do not make it easy to understand; MasterCard's policy on
interchange fees numbers over 100 pages.

Square's credit card reader was designed to bypass the fee-charging
middlemen. It has cut deals that allow it to take a total fee of 2.75 to 3
percent of each payment, which it splits with its partners, including
the credit card companies.

Square and its competitors are trying to reduce friction in the marketplace. They are trying to dial down the complication and the tens
of billions of dollars spent in the form of credit card fees, exchange
fees, or the cost of lost transactions like Jim's glass faucet. Square is
designed precisely to make commerce more fluid, allowing consumers to complete transactions without being tied to their wallets and
freeing merchants from being tied to a traditional cash register and
credit card machine.

Jack believes that Square is part of a larger trend refocusing the
economy toward bottom-up innovation. He explains, "One of the reasons we started this company, from a personal standpoint, is this trend
toward more local experiences. So I think the fabric of the neighborhood and how online is pointing to more offline local experiences is
a very, very interesting trend. You see it not just with commerce, but
you also see it with things like Foursquare, Twitter, where in the best
cases, it's an online movement that ends up in a face-to-face sort of
interaction. So in terms of what drives the economy and where the
true power lies, I think it is these networks of local sellers and local

buyers. We're seeing a shift away from the multinationals and the corporations into these local entities, and we see it certainly in our numbers." This is a dynamic we'll explore further in this chapter when we examine the sharing economy.

Square adds to that local economy by adding new capabilities for commerce onto existing technologies. "Square shows the power of distribution and distributed technologies," says Jack. "Anyone can pick up a device they already own and suddenly become a powerful commerce engine in that particular area, which then will have influence on the larger entities: the neighborhood, the city, the state, and then the nation."

Jack is also driving for Square to help combat the inequality that has proliferated alongside innovation. He makes a point of holding Square events in cities like Detroit and St. Louis that have suffered as manufacturing jobs have left the United States. He views Square as a product that can help these struggling areas incubate new businesses. "The part I think Square plays is making commerce easy," Jack says. "Not payments but commerce, so that anyone can make a start and then easily run and then easily grow." It's a commerce company for the little guy.

CODED MARKETS GO EAST AND SOUTH

In November 2012, I traveled to the West Bank during a long string of travel through the Middle East and saw where innovations like Square could be most helpful. While visiting Hebron, I encountered a joint Hamas-Fatah procession. Hundreds of men paraded down the city's main thoroughfare arm in arm, waving Fatah's flag alongside the pennant of Hamas. The warming of the historically hostile relations between Hamas and Fatah signaled a worrying renewal of militarization in Palestinian politics. Economic growth could help counter radicalization, but the Palestinian economy, especially its small business sector, is failing despite tremendous potential.

The common gripe I heard from Palestinian entrepreneurs was twofold: the lack of payment systems and the lack of access to 3G (the standard at the time for delivering wireless Internet connectivity). Each year, 2,000 Palestinians graduate from local universities in technical subjects, but only about 30 percent of them find work in their fields. As one young woman studying at Palestine Polytechnic University put it, "We must have a better economy to have better lives, and we must have 3G to have a better economy," her reasoning being that if there is no Internet connectivity, then the basic conditions necessary to do engineering work are not there. This produces a dangerous unintended consequence: underemployment for young engineers. Radicalization, unemployment, and engineering skills are a nasty combination. Many suicide bombers and bomb makers that Hamas has produced have come with this background. The best way for the West Bank to not end up like Gaza is through economic integration and well-being.

Innovations like eBay and PayPal have had a significant impact in creating the first wave of coded markets. Despite PayPal's desire to be a global payment system, there are many countries where service is limited or nonexistent. PayPal has created concern about the potential for illicit or terrorist financing, because e-transactions are less bounded to traditional systems that are routinely monitored by law enforcement and intelligence agencies. As a result, the service is banned in places including the West Bank, Pakistan, Lebanon, and Afghanistan.

But at the same time, these are precisely the places where peer-to-peer online markets and financial payments are most needed. PayPal and eBay have the potential to create new opportunities in countries and regions that are poorly served by weak financial systems.

If eBay and PayPal represent the original American manifestations of that trend, what comes next will be far more global. Among the first of these was Alibaba, headquartered in Hangzhou, China, which now employs more than 26,000 workers worldwide and operates in

48 countries throughout Asia, Africa, Europe, the Americas, and the Middle East. Its Alipay payment system performs 2.85 million transactions per minute, making it larger than PayPal or Square.

Coded markets will now reach into the world's most isolated communities, and they will link emerging markets to the global economy even more closely.

THE CODE-IFICATION OF AFRICAN ECONOMIES

Since 1998, the Congo has been mired in the globe's most lethal conflict since World War II. Driven by competition over natural resources, ethnic divisions, and the vanities of warlords, conflict in eastern Congo has claimed at least 5.4 million lives.

Although peace has been declared and elections held in recent years, the conflict continues. The current crisis has led to an almost complete breakdown of the Congolese state, which has devastated the Congolese economy. As security collapsed, most foreign investors fled—all except beer, telecoms, and mining. Without a functioning central bank, the currency has fluctuated wildly. And without access to capital or even a currency stable enough to hold value, commerce has foundered. The black market dominates the economy, and the Congo remains one of the world's poorest states. At least 75 percent of the population lives on less than one dollar a day. A third of the population is illiterate, and the average life expectancy is just 46 years.

Some of the worst of the suffering in the Congo has occurred in Goma, a city along the country's eastern border with Rwanda. Constant fighting has devastated the region and displaced more than two million people. Yet it is here that some of the most promising signs of an emerging new economy are visible, as well as signs of the remarkable potential of code-ified markets. Here, the advent of mobile payments has allowed the mobile phone to become a payment system that provides vast new opportunities for the "little guy."

In August 2009, I visited the Mugunga refugee camp located north of Goma. Hillary Clinton had been there earlier in the summer and asked me to go and see if there was some way in which we could use technology to address the problems in the region. In this shantytown lived 72,000 refugees in shacks made of plastic tarps, corrugated iron, and heavy stones to hold their makeshift constructions in place. Children walked barefoot on black-gray volcanic rock.

Even so, when I visited Mugunga, mobile phones were pervasive. The mobile penetration rate in the Congo is 44 percent. In Mugunga, cell phones were one of the few functioning parts of the economy. They were not just used to place calls. The refugees used them to send and receive money, even when they did not have a bank account. When I was there, 14 percent of camp residents had a phone, and the average number of users of each phone was three—a functional penetration rate of 42 percent. The health and living conditions were abysmal, but the residents had access to mobile phones. I would not have believed it if I had not seen it myself.

At first I couldn't understand why anyone would spend money on a mobile phone when there were so many other basic needs they lacked. A woman in the camp explained it to me: in refugee communities like Mugunga, mobile phones were how families were able to keep track of one another after they had been displaced. Previously it would have taken months or years to reunite a mother and child separated by a village raid. Now it took days or weeks. People were able to wander away from the camps for as long as they needed to find work or food without worrying that they would lose contact with their families—even as camps were built and abandoned based on the comings and goings of militias. The mobile phones also enabled residents to store what little money they had inside mobile accounts. There, safe behind a passcode, it was more difficult to steal than cash.

The pervasiveness of mobile phones could also be seen on the ramshackle houses and huts throughout Goma. Mobile phone companies like Vodafone and Tigo advertised creatively throughout the city. They paint homes for free if they are allowed to paint advertisements on the walls. Nearly all the structures were painted one of three colors, representing the three main telecommunications companies.

How did Goma and Mugunga end up with so many mobile phones? Just ten years ago, mobile phones were a rarity in Africa, but companies have rapidly tapped into this once underdeveloped market. What goes for the Congo is also true across the rest of sub-Saharan Africa. In 2002, only 3 percent of Africans used mobile phones. Today that number is over 80 percent and growing at a faster rate than any other region of the world. With few exceptions, African economies do not have legal restrictions on payment systems like those in the Palestinian Territories, Pakistan, and other places where terrorist financing is a dominant concern.

These trends are attracting the attention of some of the smartest investors and entrepreneurs from Africa and around the rest of the world. This is particularly true for those in their twenties and thirties. When people in my generation (I'm in my forties) think of doing work in Africa, charitable work or peacekeeping tend to come to mind. But I have noted that many younger people in my circle first think of sub-Saharan Africa as we thought of China and India when we were their age—as an exciting, fast-growing market worth getting into from a purely business perspective.

The man to whom many of them look for inspiration is Mo Ibrahim, a 69-year-old Sudanese entrepreneur who effectively brought the mobile phone to the continent. Mo founded Celtel in 1998, realizing that a continent of 1 billion people and almost no phones could be a highly promising market for telecom. When he started his company, the Democratic Republic of Congo had about 3,000 phones in a country of 55 million people. Today it has more than 20 million phones. Mobile telecoms spread in part because sub-Saharan Africa was a lightly regulated market. Anybody willing to invest would get the licenses, rights of way, and a market ready to buy.

As Sheel Tyle, one young investor inspired by Ibrahim, explains, before Celtel brought cell connection to Africa, "a woman who owned a salon 30 miles outside Lagos would have had to walk or find a ride into the capital every week to buy supplies herself. Now, she could make a quick phone call and have it delivered. . . . There are stories that people started moving to right around Celtel towers, like people did in the 1800s in America around railroad watering stops." Much like wells in a desert, Celtel towers built oases of opportunity that connected African communities to one another and to the wider world.

Mo Ibrahim built the network. But as with other mass infrastructure deployments, like railroads in the 19th century, the full potential of the network is borne out only when other entrepreneurs layer their creativity and commerce on top of it. How far this potential will ultimately bring Africa depends both on what new technologies and new creations will be produced and on how the financial and governance systems adapt.

Kenya's M-Pesa program is a prime example of these new technologies that show the growing power of coded money and markets. *Pesa* is Swahili for "money," and "M" stands for mobile. In communities where bank accounts are rare, M-Pesa allows customers to send and receive payments through their cell phones. In Kenya, M-Pesa has become wildly successful. By 2012, 19 million M-Pesa accounts had been created in a country of 43 million people, and approximately 25 percent of Kenya's GNP flows through the network. While estimates vary, the adoption of M-Pesa has increased rural households' incomes anywhere from 5 to 30 percent.

For such a huge impact, the way that M-Pesa works is incredibly simple. Anyone with a valid identification card or passport can register with one of the tens of thousands of M-Pesa agents in the region, located conveniently at gas stations, markets, and stores. Just hand over some cash to the agent, who then loads it onto the new account. If you want to send money, go to the M-Pesa menu on your phone and send a text to the intended recipient with the amount of money in the body of the text. Within seconds, the money is delivered. Withdrawing money is just as easy and can be done by visiting an M-Pesa agent or going to an ATM—no cards or bank offices needed. The process is safe, since M-Pesa verifies each transaction and keeps the money in an account at the Commercial Bank of Africa in Nairobi.

In addition to money transfers, M-Pesa includes loans and savings products. Safaricom recently launched M-Shwari, Swahili for "cool" or "calm," a new service that allows users to save and borrow money while earning interest. M-Shwari also facilitates the disbursement of salaries and acts as a bill-payment system. In another program, M-Pesa works with Western Union, allowing 45 countries to integrate into the M-Pesa network and facilitate international transactions.

International finance flows have a particular importance in the developing world because so much money is transferred home as remittances from workers who live abroad. Remittances are one of the primary sources of income across Africa. Approximately $40 billion is sent to communities in Africa from families abroad, and in some countries, like Lesotho, remittances comprise as much as a third of GDP. But the traditional remittance system does not work efficiently. Fees for sending remittances to Africa average 12 percent, depriving families in Africa of billions of dollars a year. And there are additional problems associated with the long distances many recipients must travel to reach remittance banks. The trips themselves are inconvenient, and to return carrying large amounts of cash makes recipients vulnerable to robbery.

Mobile remittance systems offer safer, easier, and cheaper options. They, like Square, Stripe, and Apple Pay, are all trying to reduce the friction involved in payments and money transfers. More than half the $40 billion being transferred to sub-Saharan Africa in 2015 is expected to come through mobile payments. Although most mobile payments systems are currently focused on domestic markets, companies like Kuwait-based Zain Zap are setting up mobile remittance services that promise to be far more efficient than traditional remittance banks. M-Pesa is not far behind: in 2014 it announced a partnership to set up a new system to remit money from abroad to mobile phone users in Kenya and Tanzania. Families can receive money in their own homes, with fees as low as 40 cents per transaction. In comparison, remittance fees for sending money to Kenya from the United States by traditional remittance services can run up to more than 8 percent of the total amount sent.

M-Pesa, from money transfers to savings programs to remittances, is already light-years ahead of where banking stood a generation ago—and not just in Kenya but anywhere. The year after I worked as a midnight janitor in college, I worked at a bank in Hurricane, West Virginia, spending day after day punching numbers into an adding machine, making sure that customers' monthly statements were accurate. The adding machine would produce a long, narrow piece of paper, showing the deposits and checks, which we would then stuff in an envelope and mail off. Today the notion that a human being would spend all day, every day checking math on a nonwired computer is ludicrous. The code-ification of money has changed financial services fundamentally—and done so in under a generation around the world.

Beyond greater efficiency, one of the other major effects of coded money is an increase in trust and a decrease in corruption. When I was at the State Department, Mo Ibrahim had begun to switch his focus from mobile to good governance, and I got to know him in my travels. At a dinner in London, when my deputy, Ben Scott, and I pitched Mo on our ideas for using Africa's new telecom infrastructure to encourage upward economic mobility, Mo told us that tech and telecom are all well and good, but they will matter only insofar as governance improves. Indeed, Mo had put his money where his mouth was: in 2005, he sold Celtel to a Kuwaiti company and set up the Mo Ibrahim Foundation with the proceeds. Through the foundation, he is making a significant investment in governance, focusing exclusively on rooting out corruption. Despite his background as an entrepreneur in technology and a telecom executive, Mo thinks that whether Africa takes off economically will have less to do with telecom and innovation than with whether basic governance improves. And in 2007, he created the Mo Ibrahim Prize for Achievement in African Leadership, which awards a $5 million prize and then $200,000 per year for life to African heads of state who significantly improve the social and economic conditions of their countries. In the seven years since the award was created, he has given it out only three times.

Yet there may be a way in which Ibrahim's two life pursuits coincide. Through my own work in the Congo, I saw the dramatic ways in which digital networks and trust in governance could rise hand in hand. While in Goma, my friend and then-colleague Jared Cohen and I kept hearing about the problem of corruption in the Congolese armed forces. When we researched this, we learned that the core problem was that the soldiers were not getting paid. Barrel-sized rolls of banknotes were flown from the capital city of Kinshasa to Goma, 978 miles away. After the banknotes arrived, the generals would take most of the money. Then the colonels would have their shot at it, and they might leave a little bit for lower-ranking officers. Regular soldiers would regularly go months without seeing any pay.

In response, Jared and I set up a mobile payments system so that instead of flying banknotes across the country just to be stolen by the generals, the state could transmit the funds electronically to soldiers' phones. Our thinking was that coded money could lead to coded trust and a reduction in the opportunity for corruption to come into play. The problem did not go away overnight; our team managed to get a bill through the Congolese Parliament making these transfers legal, but the generals then bribed the fund administrators to keep them from setting up the system. For years this held up the program—an important reminder that technology can go only so far in reducing corruption—but in my final year at the State Department, the system was finally put in place and the first electronic payments were successfully sent out to soldiers.

THE SHARING ECONOMY: CODED MARKETS OF TRUST

In the rise of digital payments, the issue of trust has always been just beneath the surface. Do you trust someone paying you with a beam from his or her smartphone as much as you do someone handing you cash? Do you trust the numbers showing up on your mobile phone screen as much as you trust the money in your wallet? Or, if you're an investor, do you trust that business in Africa will be safe and secure? In order to be effective, the code-ification of money, payments, and markets has also had to figure out how to code trust.

From the start, e-commerce has grappled with the question of trust—first in getting users to trust that online companies like Amazon would safely fulfill their credit card purchases, then in getting users to trust one another without ever meeting or talking to or seeing one another.

When it comes to coded trust, eBay offered the first major break-through. eBay was created in 1995, shortly after the birth of the commercial Internet, to be an online marketplace based on trust. It is a peer-to-peer network where buyers and sellers engage in commerce directly, exchanging money for goods between themselves. eBay makes its money by taking a commission fee on each transaction, and each of those transactions will occur only if buyer and seller are confident of a good outcome.

According to eBay founder Pierre Omidyar, people on eBay "learned how to trust a complete stranger. eBay's business is based on enabling someone to do business with another person, and to do that, they first have to develop some measure of trust, either in the other person or the system." The development of trust online is a product of algorithms. Despite the distance between participants, it is not blind trust. It is the exact opposite, logging reputations in a two-way rating system of buyer and seller that is monitored by the corporate owners of the platform.

My wife is a regular eBay user, and it's easy to see how effective the system is. She won't buy anything from low-rated sellers, and she always rushes to the post office within a day of making a sale so that her merchant rating stays high. She has done business with people all over America, none of whom she has met but all of whom she trusts because of algorithm-generated trust.

The next leap forward in the code-ification of trust and markets is in the so-called sharing economy. I think of the sharing economy as a way of making a market out of anything and a microentrepreneur out of anybody. The sharing economy uses a combination of technology platforms packaged as apps on mobile phones, behavioral science, and mobile phone location data to create peer-to-peer marketplaces. These marketplaces take underused assets (e.g., an empty apartment, empty seats in a car, or skill as a math tutor) and connect them with people looking for a specific service.

One of the best-known examples is Airbnb. Every time I speak with the company's cofounder and CEO, Brian Chesky, he repeats the company's creation story to me. In autumn 2007, Brian and his buddy Joe were unemployed in San Francisco and trying to figure out how to pay the rent. All the hotel rooms in the city were booked up for a conference, so they decided to use their three unused air mattresses and breakfast cooking skills to create a low-end bed and breakfast. They marketed it as "Airbed and Breakfast." It helped them pay the rent and launched the idea for Airbnb: a marketplace to connect underused lodging space with people struggling to find affordable accommodations.

Fast-forward to today: Airbnb is effectively the world's largest hotel chain without owning a single hotel room. It has more than 800,000 listings in 34,000 cities. It has housed more than 20 million people. With a valuation of $20 billion, Airbnb is worth more than twice as much as Hyatt, and Brian has gone from being unable to afford the rent to being a billionaire.

At the conclusion of Chesky's creation story, he always says, "It's like the United Nations at every kitchen table!" While the idea that

Airbnb is bringing people from around the world together is cute, it masks the economic reality. Airbnb has succeeded in bringing eBay's trust-through-algorithms-and-ratings model to lodging and built a business around it. Nobody is really sharing anything in the sharing economy. You can call it the sharing economy, but don't forget your credit card. At last measure, the estimated size of the global sharing economy was $26 billion, and it's growing fast, with some estimates projecting it will be more than 20 times larger in size by 2025. Part of why Chesky's story is cloying is that Airbnb is now a destination for castles in addition to couches. When I last checked, there were more than 600 castles available, with prices often approaching $10,000 a night. There is absolutely nothing wrong with this, but the techno-utopianism behind its origins and narrative has long been passed by economic reality. In some cases, the sharing economy has turned what might have once been a casual favor into a financial transaction. That is hardly the stuff of "sharing." In most cases, sharing-economy businesses are just businesses. Brian and Joe didn't share their spare air mattresses; they rented them out. To the extent that there is an underlying ideology, it is not about sharing or creating community around the breakfast table; it is the economic theory of neoliberalism, encouraging the free flow of goods and services in a market without government regulation.

One company that seems to recognize the sharing economy has nothing to do with sharing is Uber. Founded in 2009 by Travis Kalanick and Garrett Camp and also based in San Francisco, Uber provides travel and logistics in more than 250 cities in 58 countries as of June 2015. Uber's first tagline was "Everyone's Private Driver," but as the company has expanded, it has changed its motto to "Where Lifestyle Meets Logistics."

The impact of Uber will likely spread far beyond your nightly ride home; it has implications for business models across transport and logistics globally. Today Uber is best known for providing the equivalent of a taxi ride. But if you listen to discussions in the executive suite and with its board of directors, what you'll hear is a vision to dominate urban logistics. This starts with car rides. Uber is developing a ride-sharing model that aspires to take 1 million cars off the streets of London while creating 100,000 jobs. Even if it comes near a fraction of this goal, it is still all for the good for reducing carbon emissions and for employment.

Beyond that, expect Uber to try to take over the big business of same-day and next-day delivery. I imagine opening the Uber app on my phone when I'm looking to send a package. The app geolocates me, I'll tap a button that reads "pick up now," type the delivery address into my phone, hand the parcel over to a driver, and then forget about it. The delivery has been billed straight to my credit card. I'll be able to look at the delivery person's ratings to ensure reliability, and I'll be able to pick how quickly I want it delivered, paying a premium for the parcel to be delivered immediately. Once the package has arrived, I will rate the courier for service performance, as will the recipient. I would not be surprised to see Uber take over pizza and flower delivery services and be contracted out to pharmacies to deliver drugs to homebound patients.

At its last financing, the seven-year-old company was worth $50 billion, making it worth more than double the value of Hertz and Avis combined. High-profile investors include Google Ventures and Amazon founder Jeff Bezos.

Coded markets like eBay and Airbnb simultaneously concentrate and disperse the market. With coded markets available to even the smallest vendors, a trend has arisen that pushes economic transactions away from physical stores or hotels and toward individual people, as they connect either locally or online. This is how the market is dispersed. The route through which it is dispersed, however, redirects each of those transactions through a small number of technology platforms usually based in California or China. This is how the market is concentrated.

The power of coded markets was explained to me by Charlie Songhurst, one of the most creative thinkers at the intersection of technology, society, and the global economy. Charlie bet on Google early, after realizing the winner-take-all dynamic of online search as a young analyst at McKinsey. He then went on to work for Microsoft as head of corporate strategy, and now, at age 35, he oversees a series of his own funds. When it comes to his own lifestyle, Charlie doesn't own a car, hold a permanent residence, or employ a staff. He owns a few suitcases with personal possessions and travels the world living as part of the shared economy, relying on Uber and Airbnb.

It's fitting that while the rest of the world is just noticing the local effects of the sharing economy, Charlie is pointing out the potential for powerful global consequences: "Before Uber there was in Milan, Italy, in Lyon, France, two or three mini-cab companies that used to compete. The owner of that company would be worth 1 million or 2 million bucks. He was a rich guy in the local community. You had that in every city in Europe. They've all ceased to exist. The same thing will happen all over the world. You will still have drivers. But that's the most unskilled job in the line. The rest of the money will flow to Uber shareholders in Silicon Valley. So a huge chunk of the Italian GDP just moved to Silicon Valley. With these platforms, the Valley has become like ancient Rome. It exerts tribute from all its provinces. The tribute is the fact that it owns these platform businesses. Every classified ad in Italy used to go into a town newspaper. Now it goes to Google. Pinterest will basically replace magazine sales. Now Uber dominates transport."

He sees the same trend taking place with his permanent landlord, Airbnb, as it "will replace massive chunks of the boutique hotel industry and self-catering." On the whole, Charlie observes that as sharing platforms expand, "the value flows to one of the places in the world that can produce tech platforms. So the global regional inequality is going to be unlike anything we've ever seen."

This is an alarming trend, and to an extent, Charlie is right. There's value leaving local hubs and heading to Silicon Valley. But the drain is mitigated by a few factors. First, there is the near-inevitable fact that the large platforms in Silicon Valley will be going public. Their ownership will be much more distributed than those locally owned cab companies, and many of the beneficiaries of those early investments are pension funds that invest in the big venture capital and private equity funds. Those pension funds manage the retirement funds for people in the working class like teachers, police officers, and other civil servants. This doesn't fully account for the loss, and it doesn't negate the irony that the people driving cars for Uber don't have pensions, but it's worth noting in the face of Charlie's predictions. Also important is the fact that there is indeed new value being created in local hubs whenever platforms like Airbnb become an option.

More convincing than Chesky's "every kitchen table a UN" story are economic impact studies that suggest that most of these sharing-economy platforms are strengthening the working and middle classes. In the case of most sharing-economy platforms, the product or service being sold is a latent good—something that isn't otherwise going to be used. It is also extending economic activity into diverse communities. In New York, 82 percent of Airbnb listings are outside Manhattan's Midtown, helping to get tourist dollars into outer-borough neighborhoods such as Bedford-Stuyvesant in Brooklyn and Astoria in Queens. In the case of lodging, as with Airbnb, increased stock takes a scarce resource and makes it more abundant and therefore affordable. It lowers hotel room rates and transfers some of that value to people with a spare bedroom while also creating new value. While Airbnb rents castles and lodges the likes of Charlie Songhurst, its data also show that it is allowing people to travel who otherwise could not and is making it possible for people to stay on vacation longer. Where a typical tourist stay is three nights, the average Airbnb guest stays for five nights.

Finally, Airbnb extends the opportunity for supplemental income to hundreds of thousands of households. I think it is no coincidence that the sharing economy took off during the economic crisis, when people throughout the United States and Europe needed extra income. Half of Airbnb hosts are moderate or low income. Of the 5,600 Airbnb hosts in Berlin, 48 percent of their earnings go for essential living expenses such as rent. As many households faced the twin challenges of rising rents and the possibility of foreclosure during the financial crisis, 47 percent of Airbnb hosts say that hosting allowed them to stay in their homes.

I also think it's no coincidence that the founders of most of these platforms are from the millennial generation. Most people my age and older have to bend their mind around the idea of using a mobile app to book an overnight stay in the house of someone they've never met. For millennials, the appification of lodging, labor, and travel is more native and gives credence to the idea that the sharing economy is only in its earliest stages.

Uber and Airbnb have inspired a host of imitators, and the sharing economy is growing far beyond lodging and transport. Companies have been established to sell (not share) latent goods and services ranging from home-cooked meals and day care for pets to tutoring in math.

Imagining what is next, I think it is nearly inevitable that the sharing economy will come to include more specialized forms of labor. In the early years of its existence when eBay made anyone a retailer, the platform was dominated by low-cost trinkets and gadgets. It was basically an online garage sale. Today you can buy any make or model of Ferrari, the most precious item you might find in anyone's garage. The sharing economy started with sleeping on couches and car rides. I foresee it growing to allow workforces to be built almost entirely out of peer-to-peer marketplaces where everyone from top engineers to the janitor sells their services online, wiping out headhunters and temp agencies. And just as rare, precious items like Ferraris are now sold on eBay, I believe the sharing economy will grow to include even rarer and expensive services like surrogacy, again with user rankings establishing algorithm-derived trust for the goods and services sold.

The opportunity to work on a project-by-project basis involves trade-offs. There is more independence and flexibility but fewer worker protections and rights. This too tends to skew toward the preferences of younger workers who are less focused on entitlement programs and who don't enter the workforce expecting to have just a few employers over their lifetime.

This might be manageable if the laborer is providing very expensive, highly sought-after engineering skills, but if you are a janitor, having to migrate from a full-time employer with benefits such as workers' compensation and health insurance to brokering your services on a sharing-economy platform will lead to less well-being. When the janitor then has to list his spare bedroom on Airbnb, it is not supplemental income—it is survival income. As workers enter middle age and have kids, the need for benefits grows. If more of the labor force is sharing economy–based temporary employment without benefits, it hammers the working class and pushes them into safety-net programs. For all the efficiencies of the sharing economy, toward the end of life or if a worker becomes sick or injured, the responsibility of government increases. Worker protections have shifted from employers to taxpayer-funded government programs.

Yet as these economic changes take place, a new set of norms is being established, rooted in coded markets and algorithms. These replace the norms that have traditionally been set by government. Trust is determined by the platforms' user rankings instead of consumer protections provided by government. Trust has become code-ified, and the role of the state as a regulator has been diminished.

As the sharing economy grows as a share of the total economy, the safety net needs to grow with it. It's a necessary cost for allowing loose labor markets to work without much regulation, and if it generates enormous amounts of wealth for the platform owners, then the platform owners can and should help pay for added costs to society.

BITCOIN AND THE BLOCKCHAIN: A CASE STUDY IN CODED CURRENCY

Nothing says state sovereignty like currency. We put pictures of presidents, monarchs, and prime ministers on our banknotes. Currency has become fundamentally linked to our notion of national economies, national power, and even national identity.

Could a currency break this traditional bond with the nation-state? Could digital technology go as far as replacing banks or governments as arbiters of trust and create a new protocol for doing business around the world?

Bitcoin, a new transnational currency released in the midst of the financial crisis in 2008–2009, offers a case study for the future of currency as the code-ification of money intensifies. Bitcoin is a "digital currency"—a currency that is stored in code and traded online. It is also a "cryptocurrency," a term that is often used interchangeably with "digital currency" but signifies that the currency uses cryptographic methods in an attempt to make it secure.

Bitcoin has become the world's first cryptocurrency to gain widespread use. Although there are dozens of cryptocurrencies, it is currently the largest and most influential. At first glance, Bitcoin looks kind of like PayPal in that it offers a way to pay for goods online, with no physical interaction needed. As of the 2014 holiday season, some 21,000 merchants accepted bitcoins, including household names like Victoria's Secret, Amazon, eBay, and Kmart. At first glance, there also looks to be an investment dimension to Bitcoin. It has the properties of a speculative asset, with its value going up and down with huge gyrations. But there is much more to it.

Bitcoin has encapsulated many of the contradictions and possibilities of digital currencies in a world still largely defined by national economies and governments. Its origin comes from ideological communities deeply skeptical of governments, traditional financial institutions, and "fiat currency" (money that draws its value from government law). Bitcoin has developed a new community around an online currency trying to circumvent these established institutions.

A fiat currency fundamentally depends on trust. People must have a shared belief that the currency is ultimately worth something. The word *fiat* means "it shall be": a pronouncement from on high that the currency has value. Bitcoin came about both as a result of declining trust in the traditional financial system during the financial crisis and because of its technological advance in creating a trustable mechanism for monetary exchange online.

On October 31, 2008, a research paper, "Bitcoin: A Peer-to-Peer Electronic Cash System," was published on a cryptography listserv by a mysterious author identified as "Satoshi Nakamoto" who has kept his/her (their?) identity unknown. It called for the creation of the world's "first decentralized digital currency." Satoshi Nakamoto condemned state-based currencies:

> The root problem with conventional currency is all the trust that's required to make it work. The central bank must be trusted not to debase the currency, but the history of fiat currencies is full of breaches of that trust. Banks must be trusted to hold our money and transfer it electronically, but they lend it out in waves of credit bubbles with barely a fraction in reserve.

Bitcoin represented a different effort to rebuild trust in the financial system. In the old model, established institutions functioned as an agent of trust, protecting parties from fraud. Bitcoin comes out of a community that does not trust the old order. They seek to establish a trust-based financial system among themselves, relying on algorithms and encryption.

Just about anything can be hacked on the Internet, so the core difficulty with the creation of a digital currency is creating something that cannot be stolen or counterfeited. eBay pioneered the first breakthrough for enabling trust in the otherwise untrustworthy environment of the Internet. But for high-value transactions such as contracts and international payments, digital efforts to take out the middleman have stalled because these require the highest levels of trust. In order to achieve this supremely high level of trust, we tend to rely on banks and law firms. The boldest goals for Bitcoin go beyond its just being a currency; it could be something that creates a space for trusted transactions that have never been possible before online.

Bitcoin is more easily thought of as a public ledger system than as a physical currency. If I mine or purchase bitcoins, I don't ever receive any actual coins or tokens; instead I'm given a slot in Bitcoin's ledger. Each slot has a public address (a long string of numbers and letters saying where the slot on the ledger is) that can be used to send or receive bitcoins. And ownership—my slot in the ledger—is verified through a secret, cryptographic "private key." That private key is what my father the real estate lawyer would call a "bearer instrument"— something that can establish ownership of a property without attaching anyone's identity to it. In this way, it is different from PayPal, which identifies people through an email address or a bank account. And unlike wallets that hold cash, Bitcoin wallets do not actually house bitcoins; they store the coins' private keys. To use bitcoins in a transaction, I just need to know the public address of my coins and the public address of whoever I'm paying, and then verify my ownership of my coins by entering the private key from my wallet. Cryptographic algorithms ensure that no one else can use my funds without access to the private key, which makes it crucially important for users to keep their private keys private (and, often, stored offline).

But how can all of this actually function in a digital environment? What's to prevent me from copying my coins just as I can every other file on my computer, or counterfeiting new ones that follow the same pattern as the bitcoins already out there, or using two different devices to spend the same coins twice, simultaneously, before anyone can figure it out? How do I know that anyone I'm selling a product to actually has the bitcoins they say they do? And won't the best hackers in the world be able to break this system wide open?

Bitcoin's answer to all these questions, and its method for establishing a genuine breakthrough in digital trust, is a cryptographic invention called the blockchain. At its core, the blockchain is the big ledger on which all transactions are logged. And *every single* transaction going back to the very first Bitcoin payment is recorded on the blockchain, though they're logged anonymously or pseudo-anonymously. One of the blockchain's key characteristics is that it is public, and instead of being stored at one central location, it is distributed to every Bitcoin user. By making everything public, the block-chain reduces the possibility of fraud drastically, because you can't counterfeit the existence of property in public view. Fraud is further diminished by the fact that every bitcoin carries its history with it; to try to counterfeit a coin would require counterfeiting a false lineage going back all the way to the beginning of Bitcoin. It would never be accepted by the system, since the millions of copies of the ledger that reside throughout the rest of the Bitcoin network would not have any record of this counterfeit coin or its invented history.

A widely distributed ledger lets everyone know who has what and prevents any individual from barging in with counterfeited property. The major headache that Satoshi Nakamoto conquered, and that every previous cryptocurrency had failed to manage, was the question of how to update that decentralized ledger: How could you make sure that the millions of copies of the master ledger, which are located far and wide throughout the Bitcoin network, are all the same, all accurate, all up to date, without anyone cheating?

The answer to this question is what gives the blockchain its name and what makes decentralized digital trust actually possible. Bitcoin's software is designed so that the ledger updates at regular intervals, pooling every private-key-verified transaction that has been conducted on the network since its last update and lumping them into a big block that can be added to the ledger (the blocks, when added together, form a chain; hence the name). Nakamoto figured out a clever way to ensure that these regular updates are triggered without the need for a central authority, not even a central timekeeper. In order for a block to be added to the chain, the computers in the network first need to solve a complex, randomized, and time-consuming algorithm. Once the algorithm has been solved, the computer that found the solution then sends its solution out to the network along with the latest block of transactions that should be added to the chain. Because the algorithm is difficult to solve but easy to check, it serves as a reliable signal for telling the entire network when to update. And because the algorithm has a random element, every computer in the network has a chance at solving it, which prevents any lone powerful computer from seizing central control. The slight time buffer introduced by the algorithm also prevents any user from trying to double-spend their bitcoins, since the delay allows the network time to snuff out any attempt to use the same funds twice.

In all, the blockchain that Nakamoto invented allows Bitcoin to operate a reliable, constantly updated, and publicly available ledger of verified transactions without relying on any central figure or middleman to maintain order. It's a network of trust, created with code.

Because Bitcoin functions as a decentralized peer-to-peer digital network, there is no central bank to increase the money supply. New units of a specific digital currency are "mined" by the computers in the network. In Bitcoin, this mining is lumped together with the calculations that are used to find the next block in the chain, meaning that whenever a computer solves the algorithm to create a new block, it is rewarded with newly created bitcoins. Mining thus regulates the money supply while also offering the needed incentive for people to solve and verify the algorithms that keep the blockchain up to date. By keeping the algorithms complex—they're optimized to take, on average, ten minutes to solve—Bitcoin keeps the mining process taxing enough to regulate the steady introduction of new bitcoins while also making it very difficult to tamper with transactions.

Unlike a nation-state's currency, there is to be a finite amount of bitcoins. They will be introduced at a steady rate determined by a mathematical equation, and over time the algorithms increase in complexity while the number of bitcoins released by solving the equation decreases. The goal is for 21 million bitcoins to be mined by 2140. At that point, no more bitcoins will be introduced, and transactions will be solely based on existing bitcoins in circulation.

Both the regulation of the blockchain infrastructure and the money supply, through mining, are entirely decentralized and involve no nation-state or central bank. This is at the heart of its appeal to the cypherpunks and libertarians behind its creation and initial user base.

But what is the appeal to nonideologues, who care less about secrecy and decentralization than they do about commerce? It turns out that the benefits could be shockingly large. While many secondary elements connected to Bitcoin have been hacked, the blockchain technology itself has never been compromised. And if the surrounding infrastructure coalesces, the blockchain has the potential to make regular transactions—like the online purchases we make every day—much more reliable and much less vulnerable to fraud. Fraud protection is a built-in part of the financial world we live in, which we've simply come to accept as the cost of doing business. But Bitcoin at its best could make fraud impossible unless one's private key is stolen and make the thieves easy to find even if a key is stolen. The result could be a major drop in fraud. Furthermore, by codifying trust for high-value transactions, the blockchain could wipe out middlemen and friction in a variety of transactions, creating consumer surplus. On the global stage, it could also help bring frontier countries into the economic mainstream.

As venture capitalist Marc Andreessen describes, there is an enormous space for Bitcoin to fill: "Only about twenty countries around the world have what we would consider to be fully modern banking and payment systems; the other roughly 175 have a long way to go. As a result, many people in many countries are excluded from products and services that we in the West take for granted. Even Netflix, a completely virtual service, is only available in about forty countries. Bitcoin, as a global payment system anyone can use from anywhere at any time, can be a powerful catalyst to extend the benefits of the modern economic system to virtually everyone on the planet."

Charlie Songhurst argues that "the strength of a government's monetary system ultimately is a function of the strength of the rule of law in that country. Low-quality governments will have low-quality monetary systems. These will be the countries where Bitcoin is most likely to thrive. In the US/EU/Japan, the official currency is a fairly safe store of value (at least on a day-to-day basis) and the value of an alternative ledger system is minimal. In Argentina, Iraq, Venezuela, et al., this is not true. In those countries bitcoins will act like black-market dollars (much more useful than the official currency). But unlike black-market dollars, they can be used internationally— i.e., you can cross a border and email bitcoins to yourself, whereas dollars would get confiscated at the border." Songhurst also sees the eventual possibilities "that all currencies go digital and competition eliminates all currencies from non-effective governments. The power of friction-free transactions over the Internet will unleash the typical forces of consolidation and globalization and we will end up with six digital currencies: US dollar, euro, yen, pound, renminbi, and Bitcoin."

There are also interesting examples of how blockchain technology could increase efficiency and create consumer surplus. Marc Andreessen's partner, Chris Dixon, cites the lack of fees on the blockchain: "Let's say you sell electronics online. Profit margins in those businesses are usually under 5 percent, which means conventional 2.5 percent payment fees consume half the margin. That's money that could be reinvested in the business, passed back to consumers, or taxed by the government. Of all of those choices, handing 2.5 percent to banks to move bits around the Internet is the worst possible choice. Another challenge merchants have with payments is accepting international payments. If you are wondering why your favorite product or service isn't available in your country, the answer is often payments."

Andreessen also sees particular use in Bitcoin's seamless ability to be subdivided into micropayments. This could have positive effects on content providers, such as the newspaper industry, which has been hammered by declining subscriptions and advertising dollars. He argues, "One reason media businesses such as newspapers struggle to charge for content is because they need to charge either all (pay the entire subscription fee for all the content) or nothing (which then results in all those terrible banner ads everywhere on the web). All of a sudden, with Bitcoin, there is an economically viable way to charge arbitrarily small amounts of money per article, or per section, or per hour, or per video play, or per archive access, or per news alert."

Microcharges in bitcoins could even be used to combat spam. If it costs 0.0001 bitcoins to send an email, for instance, the effect on a regular user would be negligible. Spamming millions of email addresses with requests for urgent "travel" assistance or notifications of their "luck" in winning the Nigerian lottery may become economically unfeasible. That may be one of the best arguments for Bitcoin yet.

Others have almost quirky reasons for their enthusiasm for Bitcoin. When I asked another of the Andreessen Horowitz partners, Todd Lutwak, why he and the rest of the Andreessen Horowitz firm were such cheerleaders for Bitcoin, he replied, "Declined charges!" He pulled out his phone and showed me a dozen or so text messages from Citibank over the last couple of years alerting him that his credit card had been declined.

"This was when I was buying wine up in wine country with executives from Google. That was embarrassing!" He was visibly irritated as he showed me the text messages.

"I was a vice president at eBay, running a business that produced over a billion dollars in revenue, and my credit card got turned down all these times," he said. It's a very particular complaint, but one that taps into much broader possibilities. Lutwak believes the fraud detection devices in traditional payment systems to be awful, particularly for international transactions. He said billions on top of billions of dollars of sales are being killed through traditional fraud detection, and Bitcoin would be able to reduce declined transactions to zero.

I can understand why Todd Lutwak was embarrassed, and I can see the enormous amount of inefficiency in the payments space. The dreams of Bitcoin's biggest proponents are vast. But in order for Bitcoin to be equal to the enthusiasm coming out of Silicon Valley, there has to be a dramatic improvement in the wallets, exchanges, and payment systems surrounding Bitcoin. While the Bitcoin blockchain has never been hacked, just about everything around it has been. That is the starting point for why Bitcoin's introduction has been met with a good deal of controversy and confusion.

HACKED!

Douglas Saidenberg was living the professional life that a lot of college students imagine for themselves. He was a financial analyst at Leeds Equity, a high-flying private equity fund with offices on Park Avenue and 52nd Street in Manhattan and more than half a billion dollars under management in its latest fund. Douglas was 29 years old with a master's in finance, but his baby face and curly red hair made him look more like 19. His responsibilities at Leeds included studying new investments and monitoring existing ones in the fund's portfolio.

When his friends—and the press—began chattering about Bitcoin, Douglas decided it was time to get in. Like the diligent financial analyst that he was, he spent a month studying how to buy and store bitcoins before finally buying some. He also put a few thousand dollars of cash into an account he set up on a Bitcoin exchange called BitFloor. He then won a further few thousand dollars' worth of bitcoins on an online gambling site, Seals with Clubs. He transferred his cash and bitcoins from BitFloor to the larger and then more reputable exchange Mt. Gox. It was good timing. Just a couple of days later, on

April 17, 2013, BitFloor shut down. Feeling confident (and lucky) at this point, Douglas bought another batch of bitcoins, bringing his total to 67.3 bitcoins along with the cash in his account.

At 3:00 a.m. on May 7, the day after his last cash infusion into his Mt. Gox account, Douglas was having trouble sleeping and checked his email on the phone by his bed. He noticed a handful of emails from Mt. Gox with details of transactions that had just been made. He got a rush of adrenaline and ran to his computer. His username and password on Mt. Gox had been changed. The cash in the account had been converted to bitcoins and, together with the rest of his bitcoins, had been transferred out of his account. He had used two-factor authentication (using not just a username and password but also an authentication code sent to his mobile phone) for Seals with Clubs because he thought the site seemed a little dodgy. It had not occurred to him to do the same for Mt. Gox because it was so big and ostensibly secure.

Douglas fired off emails to Mt. Gox. It was now 3:15 a.m. He wrote:

Hello—these two withdrawal requests were not done by me! In addition the funds I just deposited were stolen from me! Can I get my funds back? How did this happen????

Three hours and eighteen minutes later, he got a reply:

Hello,

We sincerely regret for the inconvenience. Our apologies for the delay in response and we are sorry to hear of your misfortune. We do really understand that how difficult it would be to face this situation. As stated earlier unfortunately, BTC transfers are non-cancelable nor reversible. We are extremely sorry to say that we do not have an option to credit the requested BTC. Fees that you paid was for the trading service that we offered. However, we are working on new development security measures which can help our users to overcome this kind losses.

I see that the withdrawal have already been processed. Kindly refer to the below link and process withdrawal can not traced or stopped.

https://blockchain.info/address/1NRg1LwyyPGA67SqwcPkRm1e9v2mi5x 2rF

Make sure you change password immediately and do not use same username and password on other services and also make sure that the email account is secured and add a software or yubikey authentication to your account at the security center.

Please file a police report and have them contact us, and we will gladly provide any documentation for the investigation. We apologize for any inconvenience caused.

For your information, I have updated your funds to 0 and in case if you need to change the withdrawal limits you may contact us back for further assistance.

Thanks,

MtGox.com Team

At this point, Douglas was in a spin. At first, investing a few thousand dollars was not a big deal. But he had continued buying more, and the price of Bitcoin was shooting up. For a 29-year-old—even a 29-year-old working at a private equity fund—it was going to sting if he couldn't get his cash and bitcoins back.

It did not take him long to figure out that he was out of luck. "What am I going to do?" he said. "Go to the Seventeenth Police Precinct and walk up to the desk and say, 'Hey, there's this thing called a cryptocurrency, which is a virtual currency, and it was stolen from an account by anonymous hackers'?" He knew that the local cops would not even know where to begin. He filed a crime report with the FBI's Internet crime division and never heard a word back. In the meantime, the value of Bitcoin continued to rise, and the value of Douglas's loss went north of $70,000. He was out of luck.

Not long after, Doug would have lots of company in misfortune. In February 2014, hackers stole 850,000 bitcoins, then worth almost $500 million, from the Mt. Gox exchange. The company's CEO, Mark Karpeles, claimed that hackers had exploited a software bug in Mt. Gox's system, producing what is known as transaction malleability. This bug allowed hackers a short window of time to change any transaction's ID before computers in the Bitcoin network solved the algorithm to confirm the transaction. The bug has since been fixed.

It is still unclear exactly what happened and who benefited. Researchers at Swiss Federal Institute of Technology had been following occurrences of transaction malleability in Bitcoin and found 302,000 cases, but most happened after Mt. Gox's February 10 press release. These were, in all likelihood, copycats capitalizing on the initial hack. Between January 2013 and the February press release, hackers had only tried to steal 1,811 bitcoins. They were successful 25 percent of the time. This accounts for 386 bitcoins out of the 850,000 stolen. More damning evidence emerged when a document from Mt. Gox was leaked that showed the company had been losing money for years because of hackers. It is unclear whether Mt. Gox was somehow involved or was just so badly mismanaged that it was easily compromised. Regardless, Mt. Gox was forced to shut down.

As the Mt. Gox episode shows, the real security threat to Bitcoin is not the security of the blockchain but the infrastructure around it. In my professional circles, when many people think of Bitcoin, they think of hacking. And while Bitcoin advocates are technically correct when they say that the Bitcoin blockchain has never been compromised, when what you use to buy, store, and transfer bitcoins is compromised, it feels like a distinction without meaning. For all the strength of the blockchain, Bitcoin still needs an enabling ecosystem including trading platforms, payment systems, and pricing indexes to function smoothly and create the level of confidence necessary for high-trust transactions.

The attitude among the most respected investors in Silicon Valley is that the collapse of these early Bitcoin ecosystem companies is actually a good thing. They tell me that these companies had to be taken down for Bitcoin to make it over the long term.

Some people are trying novel approaches to ensuring the security of their bitcoins. One Bitcoin enthusiast from the Netherlands injected a microchip into his hand, which holds the encrypted private keys to his bitcoins. In the parlance of Silicon Valley, I don't think this is a scalable solution. It does highlight the need for cryptography literacy so that bitcoin owners understand how to keep their private keys private.

Fortunately for holders of bitcoins, other solutions are being developed. One of Silicon Valley's top minds (and nicest people), Reid Hoffman, is the founder of LinkedIn and a partner at the venture capital firm Greylock Partners. He told me about Xapo, a company that his firm invested in that addresses the security issue head-on. If the vulnerability is in hackers being able to access private keys and passwords that allow them to masquerade as the owners of the bitcoins and transfer them away from their rightful owner, then the solution, Xapo reasons, is to make private keys and passwords inaccessible. Xapo has built a network of underground vaults around the world holding confidential information like private keys and cryptographic materials. They are physically stored on servers that have never touched an outside network, including the Internet. The servers are guarded using biometrics and men with guns. Some things never change.

Other security concerns regarding Bitcoin relate to its use in the darker parts of the Internet—sites like the Silk Road and Atlantis that were used to facilitate illegal activities including prostitution, drug trafficking, and illicit weapons sales. When Bitcoin was still obscure enough to go largely unnoticed, these dark websites had a brief heyday, but law enforcement agencies have thoroughly penetrated this world and, if anything, Bitcoin has made their work easier. Although the blockchain keeps personal identities secret behind cryptographic code, in order to access the blockchain, people must leave digital footprints that law enforcement agencies know how to follow.

THE BLOCKCHAIN AND THE ESTABLISHMENT

Bitcoin initially pitted Silicon Valley against the establishment in government, on Wall Street, and among leading economists. However, much of that same establishment now sees blockchain technology as a technological solution to many high-cost transactions. Economists from the left and right, investment bankers, and government officials questioned its value and often its legality. From the right, former Federal Reserve chairman Alan Greenspan dismissed Bitcoin as a bubble, pointing out that currency "has to have intrinsic value." He went on to say, "You have to really stretch your imagination to infer what the intrinsic value of Bitcoin is. I haven't been able to do it."

From the left, Paul Krugman, the Princeton economist and *New York Times* columnist, has been even more brutal in his take on Bitcoin. The titles of his columns on Bitcoin alone give you a sense of how Krugman thinks. They have included: "Bitcoin Is Evil," "Bits and Barbarism," "Adam Smith Hates Bitcoin," and "The Antisocial Network." Krugman writes that the rise of Bitcoin is an indication that we are in a state of "monetary regress." As he describes, the economic principles of Bitcoin contradict the economic theories of both Adam Smith and John Maynard Keynes and push us back to medieval times when precious metals were the dominant store of value.

Prominent economist Nouriel Roubini sent out a string of tweets attacking the notion that Bitcoin is a currency. As Roubini tweeted: "Apart from a base 4 criminal activities, Bitcoin is not a currency as

it is not a unit of account or a means of payments or store of value."
He went on to explain his rationale in further tweets: "Bitcoin is not a
unit of account as no price of goods and services is set in Bitcoin unit
nor it ever will. So it isn't a currency." "Bitcoin isn't a store of value
as little wealth is in Bitcoin and no assets in it. Also given price volatility it is a lousy store of value." "Bitcoin isn't means of payment as
few transactions in Bitcoin. And given its volatility all who accept it
convert it right back into $/€/¥."

Roubini went even further, calling Bitcoin a scam and a fringe
movement: "So Bitcoin isn't a currency. It is btw a Ponzi game and a
conduit for criminal/illegal activities. And it isn't safe given hacking
of it." He finished by going for the economic jugular: "BTCbugs like
gold bugs are fanatics who speak of BTC in cult-like religious ways.
Like gold bugs they have paranoid conspiracy views on the $."

Many views of Bitcoin have changed as the blockchain technology has
become better understood, evolving from hostility to skepticism to a
measure of acceptance.

I could see the evolution of establishment mind-sets by listening
to Larry Summers, who has made a career working at the highest levels at the intersection of academia, finance, and government with jobs
including secretary of the treasury, president of Harvard University,
director of the US National Economic Council, and chief economist of
the World Bank, among other positions. He also has one of the sharpest minds in the world.

When I first spoke with Larry about Bitcoin in fall 2013, he told
me, "I doubt it's a profoundly important politico-economic event." He
went on to explain, "It won't be a geoeconomic event for the twenty-
first century. The more dramatic visions of stateless money seem im-
plausible because I don't think in the grand scheme of things there
is that much desire to hold gold to protect against the vicissitudes of
states. And I think that for the rest of my lifetime, gold is going to be a
better bet for people with that passion than Bitcoin."

Larry Summers is not a Luddite. He is on the boards of Square and LendingClub, a popular peer-to-peer lending platform that has processed more than $6 billion in loans. He is also an advisor to Marc Andreessen's venture capital firm.

Fast-forward 18 months, and Larry now sees the potential of the blockchain technology facilitating the "medium of exchange" property of money. He has even joined the advisory board for Xapo, the Bitcoin company Reid Hoffman funded with the underground vaults.

After an initially hostile response, Wall Street is also warming to the potential of blockchain technology. In April 2015, Goldman Sachs and Chinese investment firm IDG invested $50 million in a Bitcoin company, specifically because they liked the technical innovation that made it easy to move money around the globe.

The investment from IDG showed how governments are growing more open-minded. In December 2013, the Chinese Central Bank effectively banned Bitcoin and issued a statement about its intent to "safeguard the interests and property rights of the public, protect the legal standing of the renminbi, take precautions against the risk of money laundering and maintain financial stability." IDG's investment a year and a half later would not have happened without an eyewink from the Chinese government.

Governments have struggled to adapt their regulatory environments to Bitcoin because of the speed of its emergence, its potential for nefarious use, and the uncertainty surrounding its value and durability. As a result, many governments have found themselves in contradictory positions: cracking down on Bitcoin while laying the groundwork for its future use, confiscating bitcoins, and then, as a result, holding or selling bitcoins themselves.

The US government has had a schizophrenic approach to Bitcoin. For official US policy on Bitcoin, different agencies have adopted different, and sometimes contradictory, rules. In March 2014, the Internal Revenue Service defined bitcoins as property, not currency, subject to capital gains taxes. In its statement, the IRS noted that Bitcoin "does not have legal tender status in any jurisdiction." Yet just three months later the Federal Election Committee approved the use of bitcoins as currency for campaign donations.

In other countries, the treatment of Bitcoin has often mirrored the characteristics of the political systems and personalities in power. More authoritarian states have been quick to crack down on Bitcoin in the name of security and are seeking to minimize a potential competitor to their own control over the economy. Nations in the West have struggled to develop coherent regulations, and many have run into the same contradictions as the United States has. Many developing states have been powerless or uninterested in affecting Bitcoin at all.

Interestingly enough, Charlie Songhurst, natural contrarian that he is, says that governments themselves may soon find Bitcoin to be a useful development. In fact, the government of Canada briefly experimented with its own digital currency, the MintChip, in 2012, billed as "the evolution of currency," before scrapping the program two years later because it lacked many of the technical advances that Bitcoin uses. But as Charlie argues, it could be critical for governments to separate cryptocurrency technology from its anarchic roots: "From a government perspective, electronic transactions are much, much easier to monitor than physical transactions."

Marc Andreessen explains that "much like email, which is quite traceable, Bitcoin is pseudonymous, not anonymous. Further, every transaction in the Bitcoin network is tracked and logged forever in the Bitcoin blockchain, or permanent record, available for all to see. As a result, Bitcoin is considerably easier for law enforcement to trace than cash, gold, or diamonds." As Andreessen further describes, "Anybody who thinks Bitcoin makes it easier to do transactions that aren't tracked by the government is 100 percent wrong. The transactions all happen in public view. Anybody can look at the entire ledger and verify who owns what. So if you're a law enforcement agency or an intelligence agency, this is a much easier way to track the flow of money than cash. So I think actually law enforcement and intelligence agencies are going to wind up being pro-Bitcoin, and libertarians are going to wind up being anti-Bitcoin."

Governments, it turns out, could be the strongest rivals to Bitcoin. As Charlie Songhurst describes, "As a competitor to Bitcoin, a central bank electronic ledger system, backed by the 'full faith and credit' of a government, would achieve immediate transactional scale. This is probably the greatest threat to Bitcoin in the long term."

BLOCKCHAIN AS THE NEXT PROTOCOL

The powers that be in Silicon Valley see Bitcoin heading mainstream. But if so, where will it eventually take hold? In my view, the best case for Bitcoin is not as a currency but as a protocol, relying on the new possibilities offered by the blockchain.

In the same way HTML became the protocol markup language for the World Wide Web, the blockchain may have the technological in-genuity to become the protocol for trusted transactions. The Web was essentially made by HTML. The great innovation of Tim Berners-Lee, the Web's creator, was that he made the Internet something visible, ac-cessible, and easily navigable—and that allowed other innovations to be layered on top of the platform. The blockchain makes trusted trans-actions the basis—the protocol—on which much else can be built.

The blockchain could provide a much lower-cost solution for transactions that require a third-party intermediary as a guarantor such as legal documents, brokerage fees, and ticket purchases.

Charlie Songhurst believes that "the problem with the Internet from 1995 to 2010 was that it enabled information dissemination and communication but lacked any ability to transfer value between in-dividuals. From 1995 to 2010, every industry in information services was transformed beyond all recognition—newspapers, music, TV, etc.—as was any industry involved in communication and connection between individuals—phone, fax, auctions, recruiting, etc." Citing the lack of a mechanism for high-value trusted transactions, Charlie adds,

"Conversely, from 1995 to the present day there has been almost no impact by the Internet on the financial services or legal industries. The process of performing a wire transfer, opening a bank account, or setting up a will has remained unchanged."

Joi Ito, director of the MIT Media Lab, expands on this idea: "My hunch is that The Blockchain will be to banking, law and accountancy as The Internet was to media, commerce and advertising. It will lower costs, disintermediate many layers of business and reduce friction. As we know, one person's friction is another person's revenue."

Charlie forecasts the elimination of commissions for the sale of stocks or bonds, since they can be transferred on the ledger. He imagines that contracts can also be embedded on the ledger, including proof of ownership for hard assets like land. This is yet another example of how digital networks and digital trust can do away with traditional middlemen as arbiters and authorities.

As Charlie explained this to me, I knew he was right—that at least some of these areas are poised to transform. When my wife and I bought a home in summer 2014, the process was no different than when my parents bought a home in the 1960s. There were huge piles of paper with signatures and seals. It took weeks to sort the records, and on the day of the sale, it took hours to get through all the paperwork. The process of verification was manual and ridiculously expensive. We paid thousands of dollars in legal closing costs to verify a transfer that could be done electronically for nearly nothing if some technology-based ingenuity were applied. As I had this thought, I could not help but note that this is what my father has done for a living for 45 years. It is hard to think that a young lawyer today could count on 45 years of employment organizing legal documents for home buyers.

For all the controversy around Bitcoin's suitability as a currency, even its critics have to acknowledge how impressive an advance it is technologically (if they care to learn about it). As Marc Andreessen has described, "Bitcoin at its most fundamental level is . . . a breakthrough in computer science—one that builds on twenty years of research into cryptographic currency, and forty years of research in cryptography, by thousands of researchers around the world."

For this reason, the blockchain could endure as a platform for trusted transactions even if Bitcoin founders as currency. I can imagine investment banks establishing walled-garden blockchains of their own to save money on high-value transactions.

The necessary big change for Bitcoin when it comes to such major transactions is the use of real identity. No more anonymous accounts. This idea may be anathema to its early advocates, but the one thing that would address nearly all the problems swirling around Bitcoin is doing away with its near-religious insistence on secrecy. When land purchases are recorded in the United States, those records are public. That level of transparency may not be necessary, though it would not hurt, but at a minimum, there needs to be some way to authenticate real identity. Under this scenario, if your private key is lost or stolen, you can reassert ownership, perhaps taking advantage of advances in biometrics. Incidence of fraud and abuse would plummet. Many of the cyberlibertarians who helped give birth to Bitcoin would flee, but more mainstream institutions would feel comfortable doing business in the blockchain. The blockchain's engineering could remain decentralized, but there would need to be some multistakeholder institutions in place to help govern the blockchain, in the same way that the Internet is decentralized but has organizations running processes like domain name registration.

THE FUTURE OF CODED TRUST

The lines of the debate extend beyond Bitcoin into the larger universe of digital currencies. There are now hundreds of other cryptocurrencies, including those with names as priceless as Darkcoin, CryptoMETH, BattleCoin, and PiggyCoin. Even with all these rivals, Bitcoin's market capitalization remains well above the competition. In June 2015, it had a total valuation of $3.2 billion. Its two nearest competitors were Ripple (total value $256 million) and Litecoin ($71 million).

Many of Bitcoin's competitors seek to correct perceived shortcomings in Bitcoin itself, including its limited (and thus potentially deflationary) supply; its "irreversibility," which does not allow for errors to be corrected; and even its negative environmental impact. Mining requires significant computing power, which in turn requires significant energy. Serious bitcoin miners can spend upward of $150,000 a day on electricity nearby. Collectively, miners spend nearly $15 million in electricity each day. In 2013, the Bitcoin community had a comparable carbon footprint to Cyprus: 8.25 megatons. One solution to electric costs and overheated computers is to mine in cold weather if there is a cheap power source for electricity. A British programmer decided to build his mine in Reykjanesbaer, Iceland, so his computers can run on geothermal and hydroelectric energy while being cooled by the arctic air.

Litecoin markets itself as being more abundant and faster to mine than bitcoins. Charlie Lee, a former Google software engineer, designed Litecoin in his spare time and launched it in 2011 to complement Bitcoin. Lee said, "People like choices. You want to diversify your cryptocurrency investments." He has described Litecoin as "silver to Bitcoin's gold," and he designed the Litecoin software to produce 84 million litecoins in comparison to Satoshi Nakamoto's design for 21 million bitcoins. Lee also decided to use scrypt cryptography to reduce mining rates per unit down to 2.5 minutes in comparison to Bitcoin's 10 minutes. Lee also chose this type of cryptography, which relies on computer memory rather than processing power, to avoid the kind of high-carbon arms race he sees among miners in the Bitcoin community.

Ripple markets itself as a global payment platform by allowing members to pay in any currency—from its own currency, the ripple (XRP), to bitcoins to state-based currencies. By allowing customers to use almost any currency, XRP functions as a currency exchange and a remittance network. It has been compared to a *hawala*, a traditional Arabic exchange system used to transfer money for people who do not have access to banks, largely as a remittance system.

In contrast to bitcoins, ripples are not mined. The company simply created 100 billion ripples and put 80 billion into its account. Ripple Labs maintains the global ledger, and its servers automatically monitor the transactions to ensure against fraud. In turn, Ripple Labs plans to distribute about 50 billion of these 80 billion ripples throughout the network to reward people for building up the network. The rest will be used to fund the company. Ripple is backed by Marc Andreessen's VC firm Andreessen Horowitz and Peter Thiel's Founder's Fund.

Most Silicon Valley figures push back whenever another cryptocurrency is mentioned. Investor Chamath Paliyipatiya believes Bitcoin will continue to dominate the space. "I don't want to comment on other currencies because they're all irrelevant," he says. "It's about Bitcoin, so we should talk about Bitcoin."

Former CEO John Donahoe of eBay, one of the first companies to establish a trust-based commerce network online, said, "I don't know what Bitcoin will look like ten years from now, but I do think cryptocurrency and digital currency are growing technologies with tremendous potential. There is no reason why you shouldn't have almost perfectly secure transfer of money with traceability. Cryptocurrency and digital currency are here to stay. And it will get more powerful, not less."

So what will be the future digital currency landscape?

When I think of cryptocurrencies, I think of the search engines of the 1990s—WebCrawler, AltaVista, Lycos, Infoseek, Ask Jeeves, MSN Search, Yahoo!—and wonder if there is a Google among them. I think the vast majority of the cryptocurrencies in circulation today will disappear to nothing, but the category will endure. I think that the cryptocurrency that breaks out (whether it is Bitcoin or another) will shed its cryptolibertarian roots and embrace the responsibilities that come with being economically significant. This includes doing away with

anonymity and pseudo-anonymity. There are too many economic benefits, particularly in markets with unstable currencies and a reliance
on remittances. There are many possibilities for the blockchain technology beyond its function as a currency, and once some applications
come to market and achieve meaningful scale, people in power who
have misunderstood it or haven't realized its full potential will see
its benefits. Much as the Internet was a confusing space largely populated by technologists prior to the creation of the World Wide Web,
once the blockchain has safer and more user-friendly wallets, trading
platforms, and pricing indexes, its use will expand beyond those who
are very adept technologically.

As blockchain technology takes off, its impact will be like that of
the sharing economy and other forces of digital disintermediation: it
will force a rewrite of the compact between corporation, citizen, and
government. It is bringing frontier economies onto the global playing
field while destroying middlemen and traditional authorities.

Amid his enthusiasm, Charlie still cautions, "The unknown un-
knowns of an experiment like this are massive. We have to remember
to be humble."

FOUR

THE WEAPONIZATION OF CODE

The world has left the Cold War behind
only to enter into a Code War.

On Wednesday, August 15, 2012, a shadowy group linked to the Iranian government attacked Saudi Aramco, the world's largest energy company. Their weapon of choice: a computer virus.

In an attack that would come to be known as Shamoon and Disttrack, after words found within the program's code, hackers developed a virus that a rogue employee delivered by USB drive onto Saudi Aramco's computer network. Like an outbreak of influenza, the virus rapidly spread from computer to computer, moving from "patient zero" to a wide swath of Saudi Aramco's huge corporate network. It infected not just Saudi Aramco's main offices in Saudi Arabia but spread to workstations in several other countries, including the United States and the Netherlands.

Shamoon was designed to wipe out the memory of Saudi Aramco's computer system. Generally when a file is deleted from a computer, it can be recovered. In order to permanently delete content from hard drives, Shamoon wrote new, useless data over the original data, which prohibited the recovery of any of the infected files. Instead, when anyone tried to open an infected file, they would just see an icon of a burning American flag. For good measure, Shamoon also overwrote Saudi Aramco's Master Boot Record, which prohibited its computers from restarting.

Shamoon went beyond wiping out computer memory: it instructed infected computers to send confirmation of the corrupted hard drives back to a designated Internet Protocol (IP) address, the distinctive series of numbers designated to an individual device. The attacker was sent the IP address of infected computers as well as the number and names of files compromised. Then the list of infected computers' IP addresses was posted to the Internet as proof of the attack's success.

The virus was discovered the following day by a trio of cybersecurity firms from the United States, Russia, and Israel. In order to contain Shamoon, Saudi Aramco's entire computer network had to be temporarily shut down. All infected computers had to be replaced. It took two weeks for Saudi Aramco's network to be back to normal capacity, with people working around the clock as the virus chewed through thousands of computers a day. By the time the attack was contained, three-quarters of the company's corporate computers were affected—about 30,000 computers. Two weeks later, Shamoon also attacked RasGas, a joint venture between Qatari Petroleum and ExxonMobil.

The goal of the attack appeared to extend well beyond compromising computers. The attacks came close to disrupting energy production for the Saudis and to taking their rigs offline. Saudi Aramco is responsible for close to 90 percent of the revenue of Saudi Arabia's government. If the cyberattack had done enough damage to shut down oil production, it would have seriously damaged the Saudi economy and increased costs at the gas pump in the United States, both outcomes that would have served Iran's purposes. Saudi Arabia and the United States are Iran's sworn enemies, and abundant oil coming from Saudi Arabia helped ensure that sanctions against Iranian oil would hold. Sanctions against Iran's oil industry are especially (and justifiably) tough. Disrupting supply from the world's largest oil producer and increasing prices would have also made many countries question the virtue and value of continued sanctions against Iranian oil. Fortunately, oil production was not disrupted during the attack, but the implications of the infiltration for critical infrastructure were severe.

Saudi Aramco had been beefing up physical security since an unsuccessful terrorist attack at one of its production facilities in Abqaiq, Saudi Arabia, in 2006. Little did it know that its deeper vulnerability lay in the virtual world until Shamoon revealed the holes in its cybersecurity.

It is notable that this could happen to Saudi Aramco. Which do you think is the world's most valuable company? ExxonMobil? Apple? Not even close. Saudi Aramco is the most valuable company in the world by a long shot, with estimates beginning at $2 trillion, more than triple the market cap for Apple and seven times that of ExxonMobil as of this writing.

Business leaders around the world took notice. If a cyberattack could happen to the world's largest company operating in a secure environment, it could happen anywhere and to anyone.

Malware. Virus. Worm. Trojan horse. Distributed denial-of-service. Cyberattack. The terms for the weaponization of code are by now well known, but we are just beginning to understand their full implications.

It is perhaps ironic that one of the earliest purposes of the Internet, among certain of its developers, was to create a decentralized, distributed communications network that could survive a nuclear attack. That same distributed structure has led to a whole new class of possible attacks. And as more individuals, businesses, and governments have been incentivized to move their assets online, the weaponization of code has grown more lucrative and more destructive.

The potential for real damage on a personal or systemic level is frightening. Whether motivated by politics or profits or mayhem, the cost of cyberattacks has now eclipsed $400 billion a year, a number that is larger than the gross domestic product of about 160 of the 196 countries in the world.

As the costs of cyberattacks have risen, so has an industry devoted to countering the threat. Companies and governments are now reducing damage and costs by directing more resources to their own defense. Over the 20 years from 2000 to 2020, the cybersecurity market will have grown from a $3.5 billion market employing a few thousand people working in IT departments to a $175 billion market providing critical infrastructure to just about every kind of business, big and small.

Cyber has grown into such a mission-critical function that every Fortune 500 board chairman now should make sure he or she has a board member with cyberexpertise. A little over ten years ago, it became almost mandatory for every board of directors to have a member with expertise in auditing. In five years, any board of directors without a board member with expertise in cyber will be perceived as a shortcoming of corporate governance.

Cyber has also created a new field of confusion and tension for governments and militaries. The weaponization of code is the most significant development in warfare since the weaponization of fissile material and has created a domain of conflict with no widely held norms or rules.

TYPES OF CYBERATTACKS

The first recorded incident of hacking dates back to 1903, attributed to magician and inventor Nevil Maskelyne. John Ambrose Fleming was at the time publicly demonstrating the advances in wireless Morse code with his partner, Guglielmo Marconi, who was nearly 300 miles away. Maskelyne disrupted what was supposed to be secure wireless telegraphy to send insulting Morse code dispatches to Fleming. The message read: "There was a young fellow of Italy, who diddled the public quite prettily," referring to Marconi. A lot has changed from Maskelyne to Shamoon—from the stakes to the methodology.

There are three main types of cyberattacks today: attacks on a network's confidentiality, availability, and integrity.

Attacks that compromise confidentiality aim to steal or release secure information like credit card or social security numbers from a given system in an illicit or unauthorized manner.

The retailer Target was the victim of a confidentiality attack during the 2013 holiday season. Hackers accessed Target's payments systems and managed to steal the credit and debit card numbers of more than 40 million customers. By inserting malware—malicious software— into Target's system, hackers were able to log when a customer swiped his or her card and automatically send the information to the hackers, believed to be based in Ukraine or Russia. In addition, the hackers stole personal information—names, phone numbers, email addresses, and physical addresses—from about 70 million customers.

The hackers were never caught, and Target suffered severely for the data breach. Profits fell 46 percent in the fourth quarter of 2013 relative to the same period in 2012. The company could still face a loss of up to $420 million in legal fees, credit monitoring for customers, and payments to banks to reissue cards. It lost billions of dollars of market value, and its CEO, Gregg Steinhafel, was forced to leave the company after 30 years of employment.

The second type of cyberattack hits a network's availability— attacks typically known as denial-of-service (DoS) or distributed denial-of-service (DDoS) attacks. Denial-of-service attacks aim to bring down a network by flooding it with a massive number of requests that render the site inoperable. Distributed denial-of-service attacks are exactly the same except that the attacker has mobilized several systems for the attacks. DDoS attacks aim to use so many attackers (potentially hundreds of thousands) that it becomes nearly impossible to distinguish the attackers' traffic from legitimate traffic. This type of attack can also use hijacked systems to mask its origin. Hundreds, thousands, or hundreds of thousands of computers can be hijacked by hackers and coordinated to attack together, in what's known as a botnet.

The targets of botnet attacks are usually big corporations or governments. However, while I was at the State Department, I increasingly saw civil-society organizations and independent media organizations on the receiving end of DDoS attacks. One of the largest cyberattacks ever undertaken was against independent news sites in Hong Kong that were reporting and performing mock elections in Hong Kong during pro-democracy protests at the end of 2014. Unlike the State Department or a big company, these modestly resourced institutions often found themselves wiped offline if they provoked the wrong cyberbully, like an authoritarian government. But just as advances in cyberattacks are spreading rapidly, they are being countered by newly available advances in cybersecurity. A fast-growing cyber company, CloudFlare, recently set up Project Galileo, aimed at providing sophisticated cyberdefense to civil-society and independent media organizations that cannot afford it on their own.

Last, cyberattacks can also affect a network's integrity. These attacks are more physical in nature. They alter or destroy computer code, and their aim is normally to cause damage to hardware, infrastructure, or real-world systems. Once an integrity attack has taken over a machine, the machine ends up being rendered useless and is added to the waste stream. The Shamoon attack was an example of an integrity attack.

Attacks can be blended. A small-scale confidentiality attack with the goal of compromising the integrity of a media company's network occurred in April 2013, just after I left the State Department, when an old adversary of mine, the Syrian Electronic Army, hacked into the Twitter account of the Associated Press. Just after 1:00 p.m., the AP's

Twitter feed posted this story: "Breaking: Two Explosions in the White House and Barack Obama Is Injured." From 1:08 p.m. to 1:10 p.m., the New York Stock Exchange plummeted 150 points, devaluing the market by about $136 billion. When it became clear that the information was false, the market bounced back to its prior level.

How did the Syrian Electronic Army do it? They used a simple computer technique called phishing. To phish, a hacker typically sends an email that comes from what looks like a legitimate source. The email contains a link asking the person who received the email to enter personal information on a website that also looks legitimate. Once the personal information is obtained, the hacker can launch spyware against the email recipient's network. The term *phishing* originated from the idea that the hacker is fishing with the bait of an innocent-looking email for sensitive information.

In the case of the AP hack, a staffer received an email from what appeared to be another AP staffer with a link to a *Washington Post* article. When the staffer clicked on the link and logged into the website, the hackers had what they needed to launch the attack. It was a simple confidentiality hack that then transitioned into an integrity attack that moved the market in a big way.

AIR, LAND, SEA, SPACE, AND CYBERSPACE

In light of the damage that even simple cyberattacks can cause, most countries are developing cyberdefense strategies. The most sophisticated governments in the cyberdomain are the United States, China, Russia, Israel, Iran, and the United Kingdom, each with different motives and boundaries for what is and is not allowable behavior.

While I was at the State Department in 2011, Secretary of Defense Robert Gates formally declared cyber a domain of warfare, alongside air, land, sea, and space, and President Obama declared America's digital infrastructure a "strategic national asset," which created the legal cover for a wide variety of offensive and defensive measures to be enacted at the newly created US Cyber Command, as well as in other parts of the US government. It's a domain that the United States takes as seriously as the more conventional modes of warfare. For example, when the United States was planning the first attacks against Libya's air defenses in 2011, one of the questions was whether we could take out their defenses with a cyberattack instead of bombs, which could create more collateral fatalities.

Notably, one of the things that was not allowed in any of the new regulations was anything approximating corporate espionage. It would remain illegal to steal trade secrets of any foreign companies and hand those over to American companies. With this policy, the United States's attitude toward cyber largely mirrors that of the British and the Israelis. Israel is a little more aggressive and Her Majesty's Government a little less aggressive. But for all three, security—the idea that they are protecting the homeland or the lives and territory of friendly governments—is what governs activity. Nations like China have been more aggressive when it comes to private enterprise.

China has been moving full steam ahead to develop its cybermilitary capabilities since the late 1990s. At first, it experimented only with communication interruptions. But since the early 2000s, China has been pursuing a deliberate strategy of cyberespionage. Simple attacks against Taiwanese and South Korean official websites have given way to more sophisticated ones. In 2002, the Dalai Lama was thought to be the victim of a Chinese Trojan horse attack wherein a virus masquerades as harmless in programs like Microsoft Word but sends private information to another entity.

China's most powerful cyberattacks have been rooted in corporate espionage: stealing intellectual property and trade secrets that it can use to help its state-owned and state-supported enterprises. Its cyberpolicies cohere with the rest of its military and economic policies that aggressively assist Chinese corporations, even at the cost of angering the rest of the world.

An American group of powerhouse former corporate, military, and government officials led a commission that published findings in May 2013 estimating that the annual losses to intellectual property theft from China exceed $300 billion, comparable to the total amount of goods exported annually from the United States to all of Asia. The NSA director at the time, Keith Alexander, has estimated the total value of all American intellectual property at $5 trillion, of which China is stealing 6 percent each year.

Of course, the United States is not the only country on the receiving end of these IP thefts. The Canadian telecommunications company Nortel Networks, which at its height employed 94,000 employees, went bankrupt in no small part because of a decade of cyberespionage. From 2000 to 2009, when Nortel Networks filed for bankruptcy, the company was hacked repeatedly by Chinese agents, losing much of its intellectual property to Chinese companies that sold similar products.

Many countries, the United States included, engage in cyberactivity that raises the ire of other countries around the world, but when it comes to the mass theft of intellectual property, China has no peer.

In 2013, an American cybersecurity company released a report that detailed the scale and scope of Chinese cybercapabilities, with a focus on People's Liberation Army (PLA) Unit 61398. Working out of a location in Shanghai's Pudong District, PLA Unit 61398 is thought to be the most advanced and best funded of the 20 cybermilitary units in China. Since 2006, the unit has been responsible for attacks against companies, mostly American, in almost every sector, including IT, transportation, financial services, health care, education, energy, and mining.

In May 2014, the US Justice Department indicted five officers from PLA Unit 61398 in response to several attacks that had taken place against heavy industrial companies in the United States, including Alcoa and United States Steel Corp. The language of the indictment states the allegations plainly:

The indictment alleges that the defendants conspired to hack into American entities, to maintain unauthorized access to their computers and to steal information from those entities that would be useful to their competitors in China, including state-owned enterprises (SOEs). In some cases, it alleges, the conspirators stole trade secrets that would have been particularly beneficial to Chinese companies at the time they were stolen. In other cases, it alleges, the conspirators also stole sensitive, internal communications that would provide a competitor, or an adversary in litigation, with insight into the strategy and vulnerabilities of the American entity.

In response to hacking allegations, the Chinese consistently deny everything, despite the extensive record collected by the US government and others, much of which is now public because of the Department of Justice indictments. The Chinese never concede anything in public and only very rarely do so in private. "The Chinese government and Chinese military as well as relevant personnel have never engaged and never participated in so-called cybertheft of trade secrets," said Foreign Ministry spokesman Hong Lei after the PLA 61398 indictments.

Yet as much as the Chinese steal, they have shown little inclination to this point to undertake integrity attacks that would have destabilizing effects on markets. As the world's second largest economy, China is now a major equity holder in just about every type of financial asset: they are invested in stability and economic growth. Their cyberincursions give them the capacity to break things, but they would just be hurting themselves if subways went dark, if corporate records were leaked online, or if any other type of headline-grabbing cyberattack took place around the world. The Chinese are taking advantage of the gray area that surrounds cyber in the international arena. They are able to engage in cyberattacks without triggering extensive sanctions or punishments for treaty violations. They have a financial and political stake in things staying calm; they have incentives to steal but not to break.

Other governments have been quicker than the Chinese to use cyberweapons to destroy or disrupt the systems of foreign companies or countries. Iran's Shamoon attack serves as an example, and another can be seen with the Hermit Kingdom.

North Korea is a hellhole of a country, with a per capita GDP of less than $2,000—lower than Yemen, Tajikistan, and Chad and about one-sixteenth the size of South Korea's GDP. The government derives its power through the twin pillars of state repression and an all-encompassing propaganda apparatus. But it manages an active corps of state-funded hackers, spending an estimated one-third of its GDP on military activities, including cyber.

The FBI attributed a successful cyberattack against Sony in December 2014 to North Korea after the nation's leaders objected to a comedy produced by Sony's film division that revolved around a plot to assassinate Kim Jong Un (Lil' Kim, as he is known to the State Department's Asia hands). A spokesman for North Korea's Foreign Ministry referred to the film as "the most undisguised terrorism and a war action." The FBI alleges that North Korea then engaged in a confidentiality attack against Sony Pictures, illicitly gaining access to troves of internal communications and data and releasing them on the web.

Then something interesting happened. President Obama promised a "proportionate response" to the Sony hacks and said that he had conferred with the Chinese. Two days later, the Internet in North Korea (a small network strictly limited to the country's elites) went dark. Notably, all of North Korea's Internet connectivity is supplied by a single Chinese company, China Unicom. I believe the White House asked the Chinese to punish the North Koreans. The Chinese would have done so because they would have been angry at their satellite state acting in a destabilizing way and without their blessing. By taking North Korea offline, they were reminding the North Koreans who controls their networks, and they were doing a favor for the United States that served their own purposes.

CYBERATTACKING EVERYTHING

As the Internet grows, it is expanding not simply to new users but to entirely new devices, well beyond standard computers, tablets, and smartphones. Electronic communications and electronic sensors have been around for some time, but the costs of sensors and data storage have recently plummeted—in part due to cloud computing. As a result, the stage is now set for what has become known as the "Internet of Things," where any object has the potential to transmit and receive data, from cars and farm equipment to watches and appliances, even clothing.

The digitization of nearly everything is poised to be one of the most consequential economic developments of the next ten years. Cisco Systems chairman John Chambers has said, "We will look back one decade from today [2014] and you'll look at the impact of the Internet of Everything, and I predict it will be five to ten times more impactful in one decade than the whole Internet to date has been." From 2015 to 2020, the number of wireless connected devices is going to grow from an estimated 16 billion to 40 billion. Chambers predicts that the Internet of Things will grow to be a $19 trillion global market. For context, the GDP of the entire world is currently just a little more than $100 trillion.

The growth of the Internet of Things is motivated by four main drivers. The first is the number of Internet-connected cars on the road, expected to grow from 23 million in 2015 to 152 million in 2020. The second driver is the advent of wearable technology, which doubled in use between 2013 and 2014. The third driver is the addition of smart controls in our homes, from thermostats to security systems to just about everything else. According to a Juniper research report, revenues generated from smart home services are expected to reach a global market value of $71 billion by 2018. The fourth driver is in manufacturing. A McKinsey report projects that by 2025, Internet of Things applications could have an economic impact of $900 billion to $2.3 trillion a year in manufacturing alone. McKinsey bases this estimate on potential savings of 2.5 to 5 percent in operating costs, the integration of the Internet of Things into the power grid, and its applications in public-sector services like waste, heating, and water systems that they believe could cut waste by 10 to 20 percent annually.

There's one huge catch: with the rapid growth of these technologies, we are also creating an almost unimaginable new set of vulnerabilities and openings for cybersecurity hacks. As the Internet of Things is rising, cybersecurity has not kept pace. "Security has often been an afterthought in the design of those systems," says Chris Bronk, a computer and information systems professor at the University of Houston.

The breach of confidentiality that occurred at Target was in many ways a precursor to what's possible in a world that's connected by an Internet of Things. In the Target hack, the tens of millions of credit card records were accessed because of a hack on Fazio Mechanical, a small company in Sharpsburg, Pennsylvania, that does heating, air-conditioning, and refrigeration jobs. The hackers stole the network credentials that Target had given to Fazio Mechanical as a vendor, and they used those to tunnel through Target's systems all the way to the point where they could reach the system that unites Target's many point-of-sale consoles (the devices through which you swipe your credit card when you're checking out in a Target). Because all of these were connected, the hackers were able to install card-logging malware en masse on the point-of-sale consoles. Target is a company with a market capitalization of more than $50 billion and 347,000 employees, but its size and connectivity ultimately made the damage worse: one simple hack from afar compromised tens of millions of credit cards.

To illustrate one of the worst examples of what could happen with a hack on the Internet of Things, X-Lab director Sascha Meinrath looks at pacemakers: "Everyone talks about the benefits of connecting to the cloud. But that assumes that the cloud will be secure. . . . People are talking about pacemakers hooked up to the cloud. There's a benefit to that—it could automatically shock you if it senses something is wrong. But what if a terrorist, or a kid doing a prank, decides to shock all the pacemakers in America?"

As he was explaining this to me, I was imagining what could happen if a system controlling home care robots were hacked. Could it be a way to hurt people? In July 2015, hackers managed to remotely infiltrate and shut down a Jeep Cherokee while it was speeding along the highway. What if, 20 years from now when some variant of the Google car has taken over the highways, someone were to hack the entire network of Google cars? Imagine a highway's worth of connected cars all going haywire at the same time—the potential for a pileup bigger than anything we've ever seen.

Soon all the networked "things" in our lives could potentially be used as hacking platforms. It's hard to imagine your refrigerator being hacked, but the reality is that it has already happened.

In January 2014, security provider Proofpoint uncovered a phishing attack that targeted consumer devices including home routers, televisions, and, yes, refrigerators. The company said in a statement, "Just as personal computers can be unknowingly compromised to form robot-like botnets that can be used to launch large-scale cyberattacks, Proofpoint's findings reveal that cyber criminals have begun to commandeer home routers, smart appliances and other components of the Internet of Things and transform them into 'thingbots' to carry out the same type of malicious activity."

What kind of malicious activity? For one, the computing power of these thingbots could be harnessed for DDoS attacks and as spam engines—anything that requires raw, dumb computing power. Mikko Hypponen, a Finnish cybersecurity expert and chief technology officer at F-Secure, a Finnish antivirus and security firm, says that another use for these kinds of "thingbots" could be to mine cryptocurrencies.

"Why would anybody hack a toaster? Why would anybody hack a fridge? A toaster or a fridge doesn't have a traditional user from which you could steal anything at all, but they do have computing power, and they are online," says Hypponen. "I am forecasting that we will see toaster botnets—infected toasters, or something like that, you know, appliances, which it doesn't seem to be any logical reason why anybody would infect them, but we will see them getting infected for their computing power to mine cryptocurrencies of the future. That's going to happen."

All of the exciting possibilities that come with connecting our "things" to the Internet are also accompanied by dangers. When I was growing up in West Virginia, it was considered smart to cancel home delivery of your newspaper when leaving for vacation so that burglars would not see a stack of papers in front of your house and know that it was a prime candidate for robbery. With connected homes, today's smart burglar is able to hack inside a smart home network and monitor when people are home. They can gather precise data about the comings and goings of everybody living in the home. If there is a security system, they can turn it off. The streamlined systems that make everyday life easier for you can, if put in the wrong hands, also be used to make life a whole lot worse.

CLANDESTINE OPERATIONS: EVEN THE SPOOKS ARE SCARED

The emergence of cybercrime has created a new imperative for government to protect its critical infrastructure and its citizens. To explain how the boundaries of law enforcement and defense are expanding, I sought out Jim Gosler, the closest thing the CIA has ever had to a cyber guru. Gosler has a kind, grandfather-like demeanor, complete with a long white beard. A retired navy captain and now a senior fellow at the Johns Hopkins Applied Physics Laboratory, Gosler founded the CIA's clandestine operations unit in its information technology division. That means he was constantly on the lookout for threats to the government's and the military's infrastructure.

Gosler has earned numerous awards: the CIA's National Intelligence Medal of Achievement, the DONOVAN Award, the Intelligence Medal of Merit, the CIA Director's Award, and the Clandestine Service

Medallion. He has also earned the Legion of Merit for exceptional meritorious performance, one of only two US military decorations awarded as a neck order. The other is the Medal of Honor—the highest award given by the US government. He is the most decorated person to ever work in cyber in the American intelligence community.

In the lobby of the Marriott Key Bridge in Rosslyn, Virginia—the epitome of the kind of nondescript suburban hotel where these CIA guys like to meet—Gosler shares his fears of the potential for cyber-attacks. While the rest of us enjoy technology (e-commerce, online banking, Uber), Gosler looks for the vulnerabilities: "What makes cyber so potentially devastating is first and foremost our utter dependence on the stuff for everything that we do in life. It's easy to grasp and understand the benefits [of digital technology]. It's not so easy to understand our dependence on it and consequences associated with being denied that stuff, based on the unbelievable dependency that we have. Medications, banking, medical, just information you know. . . . I tend to think more in terms of things from an intelligence and military perspective, but an average Joe's way of life would be dra-matically affected."

He cites the example of GPS and navigation. People in the navy once knew how to navigate by looking at the stars to determine where they are and what course to set. "Nobody knows how to do that now," says Gosler. Our navy is bound to GPS to navigate the world, and so is just about everyone who owns a smartphone. But Gosler says that GPS is hackable. The damage could be benign—getting lost driving to a meeting—or it could be catastrophic. Imagine, for instance, if some-one were able to hijack GPS systems and then direct military units on patrol toward enemy positions.

Gosler understands, as few others do, the potential for things to go wrong across the industries of the future. Ultimately, Gosler be-lieves that the prevention of the worst kind of cyberattacks, like those against power plants or air traffic control systems, falls into the do-main of government. He believes that democratic governments need to both recruit the right people and foster a close relationship between the public and private sectors to guard against major threats. With the weaponization of code evolving to include infrastructure connected through the Internet of Things, Gosler calls for us to recruit, train, and mobilize a new force of cyberwarriors.

"There are about one thousand security people in the US who have the specialized security skills to operate effectively in cyberspace. We need ten thousand to thirty thousand," Gosler says. The government's relative lack of top cybersecurity personnel is exacerbated by the choices that people with serious cyberskills have as they enter the workforce. For a nuclear scientist, the choice is to either become an academic or go to work for government. For a computer scientist with cyberspecialization, government has to compete with higher-paying private sector jobs. Cybersecurity professionals in the United States already have an average annual salary of $116,000, nearly triple the average median income. Protecting corporate networks is better paying than protecting the GPS system.

Yet Gosler is emphatic about the far-reaching importance of national cybersecurity: "You have to assume that we have the greatest dependence on this stuff. Maybe Europe is on par with us, but the scale of the United States is so much bigger. Plus, there is so much economic entanglement in the world today that if there's a massive failure of US banking, that just doesn't affect the United States. Or if there was a massive impact in European or Japanese banking, the global ripple effect of all that is pretty significant." Gosler's view reflects the Wild West–like atmosphere of global cyberpolicy, which lets China and other powers take advantage of the lack of central authorities or treaties that could dial back actions like the theft of corporate data.

Gosler's is not a lone voice. James R. Clapper, director of national intelligence, warned Congress in February 2015 that cyberattacks pose a greater long-term threat to national security than terrorism. Gosler wants the United States, with the government in the lead role and the private sector supporting it, to take on a massive effort to provide cyberdefense that protects the United States but also much of the rest of the world.

When even senior ex-CIA like Jim Gosler show fear, it's clear that the weaponization of code is real and serious. His fears come from experience—from working in windowless rooms with access to the most sensitive information. And his tone is the same as most of those who have worked at the highest ranks of government. As a rule of thumb, the higher-ranking the person in government, the more apocalyptic his or her language about cyber.

This is a major contrast with most of my friends from Silicon Valley, who view US government officials as too dark. My Silicon Valley friends are more technologically optimistic than the military, CIA, and diplomatic officials, but that's because they have not sat in the White House Situation Room and been witness to some of what has been averted. They know about the hacks against Sony, Saudi Aramco, and Target, and they know they have to beef up their own cyberdefenses, but most of them don't know what we have missed.

They know that China is stealing their intellectual property, but they have not contemplated what it would mean if China's net behavior started looking more like Russia's.

RUSSIA INVADING WITH BOTS AND BOOTS

As protests stormed through the Ukrainian capital of Kiev at the beginning of 2014, the United States and European nations were watching closely to see if Russian troops were mobilizing at the Ukrainian border, primed for intervention or invasion. But long before the Ukrainian president, Viktor Yanukovych, was ousted and Russian troops seized the Crimea, Russia had already attacked, and not by any of the traditional domains of warfare of land, sea, or air. It attacked in cyberspace.

Ukrainian computer networks years earlier had been infected by the cyberespionage package Ouroboros, named for the Greek mythological serpent pictured eating its own tail. The malware was "designed to covertly install a backdoor on a compromised system, hide the presence of its components, provide a communication mechanism with its [command and control] servers, and enable an effective data exfiltration mechanism." Ouroboros provided its developers the ability to surveil and extract information, and it established a beachhead for a future attack on infected systems.

As protests in Ukraine escalated in 2014, so did the malicious activity on Ukrainian computers. Suddenly Ouroboros was springing to life. Evidence, such as the time zone in which the malware developers operated (Moscow's) and snippets of Russian text in the code, among other factors, suggested that the Ouroboros operation originated in Russia. And when tensions between Ukraine and Russia erupted, malicious cyberactivity between the countries also spiked.

Tracking this kind of activity requires looking for malware callbacks, which are basically the communications sent from infected or compromised computers back to the command-and-control server of the attacker. FireEye, a global network security company that analyzes millions of these communications annually, tracked the evolution of malware callbacks and found a correlation between the overall number of callbacks from Ukraine to Russia and "the intensification of the crisis between the two nations."

During this time, I was hearing from people on both the Russian and Ukrainian sides of the conflict making allegations about the damage caused by the other. I had some personal history with Ukraine during my time at the State Department. My great-grandfather was born in Kiev and became an anarchist, back when anarchism was an actual political movement, before coming to the United States to escape the Russian czar's police. The Ukrainian hacker community loved that piece of my family history, as well as the dust-ups my State Department colleagues and I had with the Putin government in Russia and the pro-Russia politicians in Ukraine. After the protests began, I was banned from Ukraine. One pro-Russia member of parliament justified banning me from the country by saying that I was "the world's best specialist in organizing revolutions through social networks." The charge was flattering but gave me too much credit.

Even after the pro-Russian leader of Ukraine resigned and fled the country, the cyberattacks continued. Shortly before the Ukrainian presidential election in May 2014, the Ukrainian Security Service (SBU) announced that it had arrested a group of pro-Russian hackers who tried to disrupt the election results. According to SBU head Valentyn Nalyvaichenko, the hackers had compromised the main servers of the Central Election Commission website and had planned to destroy the election results and replace them with their own fabricated results. As SBU was foiling the attack on the night of the election, Russia's state-owned Channel One reported that far right–sector candidate Dmitro Yarosh, who had received less than 1 percent of the vote, was leading the election with 37 percent of the vote, showing a screenshot of the Central Election Commission website. The screenshot was from the hacked site, and SBU noted that "offenders were trying by means of previously installed software to fake election results in the given region and in such a way to discredit general results of elections of the President of Ukraine." Pro-Russian hacktivist group CyberBerkut ultimately took credit for the hack, and Channel One did its job misinforming and manipulating the public.

The Ukrainian conflict was not the first time that Russian hackers took to the web during a political dispute between Russia and a former Soviet state. Both Estonia and Georgia have previously faced cyberattacks from Russia.

In 2007, Estonia relocated a controversial Soviet war memorial, the Bronze Soldier, from the center of the capital city of Tallinn to a military cemetery. To many Estonians, the monument, which honored Red Army liberators, was a symbol of Soviet occupation following World War II. To Estonia's Russian community, the monument symbolized the Soviet victory over Nazi Germany. The move by Estonian authorities riled the Russian Federation and triggered a series of cyberattacks on Estonian government websites, bank websites, and media sites from April through May 2007. The wave of denial-of-service attacks took down the websites of two major Estonian banks, all of Estonia's government ministries, and several political parties for about ten days.

Estonia's foreign minister, Urmas Paet, accused the Kremlin of direct involvement. Eventually a Kremlin-backed patriotic youth group, Nashi ("Ours"), took responsibility for the series of DoS attacks: "We taught the Estonian regime the lesson that if they act illegally, we will respond in an adequate way," said Konstantin Goloskokov, a commissar in the youth group. "We did not do anything illegal. We just visited the various Internet sites, over and over, and they stopped working."

A year after the Nashi cyberattacks against Estonia, cyberattacks were used in concert with traditional military operations for the first time. Just before Russian tanks rolled into Georgia in August 2008, botnets were already on the offensive, inundating Georgian government websites with an unmanageable flood of traffic. Like the Estonian attacks in 2007, the operation included a wave of denial-of-service attacks but also the defacement of several public websites. The websites of the president of Georgia and the Georgian Ministry of Foreign Affairs were replaced by collaged images of Georgian president Mikheil Saakashvili and Adolf Hitler. The website of the National Bank of Georgia was defaced and replaced with photos of 20th-century dictators side by side with Saakashvili. The attacks continued throughout the conflict until a cease-fire agreement was signed.

From Ukraine to Georgia and Estonia, Russia's use of cyberattacks is eye-opening. It illustrates how our definitions of combat and warfare are changing as nations exchange blows in the virtual world ahead of or even instead of actual armed conflict.

COLD WAR TO CODE WAR

As long as there has been war, societies have sought to mitigate its effects. They have sought to limit its frequency, scope, and methods.

Throughout history, religious, moral, or chivalric codes have imposed limits on war. And in the past century, the concept of international law was developed to restrict states from attacking others. Societies have sought to create clear distinctions between combatants and non-combatants, between the battlefield and the home front, between a just war and an unjust war.

Cybercombat is a distinctly 21st-century form of conflict, and the norms and laws that were developed in prior centuries simply do not apply. The weaponization of code is the most significant development in warfare since the development of nuclear weapons, and its rapid rise has created a domain of conflict with no widely accepted norms or rules. Some nations are working to find a way of creating rules for the global community to abide by, but there are vast distances between stakeholder groups and therefore little hope of even a modest agreement of any sort.

The analogy that most foreign policy hands point to as a possible precedent for containing cyberweapons is nuclear nonproliferation: the creation of arms control agreements, treaties, United Nations resolutions, and international monitoring programs to govern the spread and use of nuclear weapons. Under this international framework, nuclear war is still a threat, but nuclear weapons are well understood and there are processes in place to manage them. In the 20th century, similar sets of procedures and rules were also developed for the weaponization of airplanes, space, and chemical and biological weapons.

But the confounding factor when it comes to cyberwar is that the barriers to entry are so much lower in cyber than in any of these other domains. Any country, or even any rogue group or individual, that puts a little bit of time and effort into it can develop some nasty offensive cyber capabilities. It is, in fact, the near-opposite of the development of nuclear arms, which requires years of work, billions of dollars, and access to the scarcest of scarce scientific talent and transuranium elements.

To create a cyberweapon, all one needs are a computer, an Internet connection, and some coding skills. The development of cyberweapons is incredibly difficult to trace. And as Jim Gosler observed, the nonphysical nature of cyberconflict has also made the private sector a combatant. Since national borders have much less meaning online, there's little to stop hackers from going right for the valuable assets. Increasingly cyberattacks are directed from a country toward a company and from a company toward a country.

One of the most notable cyberattacks that took place (and went public) during my time at the State Department was a Chinese government operation that targeted 34 American companies including Google and some of America's biggest defense companies. Afterward, executives from those companies came to Washington to press the Obama administration to make a big diplomatic issue of the cyberattacks. The attack helped lead to a series of executive orders and other administrative actions that elevated cyber from being something at the edges of our foreign policy to the foreground.

This won't always be the way companies respond. It is only a matter of time before some hotshot group of engineers recognizes and stalls a cyberattack and, instead of calling law enforcement or some other part of government, launches a counterattack against the aggressor. I wonder what would have happened if, when Google had identified the source of the hack, it had responded in kind with an attack designed to disable its attacker's network and computers. The Google engineers are some of the best in the world. Would China have considered this an attack or some other form of invasion? It might have.

To complicate matters even further, the layout of the Internet scrambles the traditional idea that both sovereign countries and warfare are tied to geography and physical place. A company may be headquartered in one country but have networks and servers in another. If those networks and servers are attacked, is it the responsibility of the headquarters country or the country where the servers are located to respond? If neither government responds and the corporation defends its network with a cyberattack of its own, who else does this entangle? If international norms and treaties are not agreed to, setting definitions and boundaries for cyberconflict, a cyberwar is just as likely to be fought between a country and a company as it is between two countries.

This blurring of lines calls into question the role of government and its responsibility to protect its citizens and corporations. Throughout summer and fall 2014, the Obama administration monitored a hack of JPMorgan Chase and other American banking institutions as a national security threat that required the direct engagement of the president of the United States.

For hundreds of years, the way a bank robbery took place was by people entering a bank with guns and leaving with other people's money. It then became the responsibility of government's law enforcement sector to find, arrest, and discipline the thieves. The question being asked inside the White House Situation Room today is whether government should treat a cyberattack draining bank accounts of an American bank on American territory as an attack against the American nation, as a robbery, or as something else entirely.

There is potential for things to get even freakier if you start contemplating how the Internet of Things offers a platform for both attacks and surveillance. " 'Transformational' is an overused word, but I do believe it properly applies to these technologies," former CIA director David Petraeus said of the Internet of Things, "particularly to their effect on clandestine tradecraft. Items of interest will be located, identified, monitored, and remotely controlled through technologies such as radio-frequency identification, sensor networks, tiny embedded servers, and energy harvesters—all connected to the next-generation Internet using abundant, low-cost, and high-power computing."

Petraeus's remarks came just before revelations of widespread US government surveillance surfaced in 2013, sparking international debate about the line between national security and information privacy. The revelations showed the already existing capabilities of the National Security Agency (NSA) in mining phone and email data, and they cast Petraeus's comments in a more ominous light. Imagine how all the advances of the Internet of Things will also bring new concerns about privacy. If your garage door knows when you're getting home from the airport, so might a government surveillance program. If your watch is not just telling time but your location, schedule, and communications, that makes it a device that is ripe for the hacking.

The Cold War did not lack for political and military tension, but it did have a clear set of alliances organized around the binary of the struggle between Communist and Western bloc countries. The Code War has no such simple organization, and traditional alliances have fractured. After the revelations of Edward Snowden, the governments and public of European countries condemned American cyberpractices. Billions of dollars of business were lost by American telecommunications and technology companies, which were no longer trusted. One study pegged the loss to American businesses in the cloud computing industry alone at between $22 billion and $25 billion over three years.

Yet there is little to no prospect for any sort of short-term progress to be made developing international law, treaties, or other frameworks establishing norms and rules for cyberactivity. The United States won't agree to anything that the Europeans would demand that limits intelligence-gathering activities. The Chinese won't admit to, much less agree to, anything related to industrial espionage. The Russians have gone on the attack. And the nonstate actors that supply much of the conflict in the cyberdomain will never accede to the niceties of agreements forged by governments.

In the face of this unhappy reality, the American government is increasingly turning to its private sector as a partner. In February 2015, President Obama signed an executive order making it easier for government and business to share information about cyberattacks and collaborate on countermeasures. The army has gone so far as to publicly release forensic analysis code called "Dshell" onto the software collaboration platform GitHub. Justifying the unusual step, William

Glodek, network security branch chief of the US Army Research Laboratory, said, "Outside of government, there are a wide variety of cyber threats that are similar to what we face here. . . . Dshell can help facilitate the transition of knowledge and understanding in academia and industry who face the same problems."

The Dshell program is laudable but falls well short of the real changes needed in international law, treaties, or other frameworks for establishing norms and rules for cyberactivity. It will take a cyberattack that produces a large body count or something that has a negative impact on the GDP of countries on all sides of a cyberwar to get the United States, China, and Russia to agree to anything meaningful. Until then, the cyberdomain will remain the Wild West.

THE CYBER-INDUSTRIAL COMPLEX: THE WEAPONIZATION OF CODE AS AN INDUSTRY OF THE FUTURE

The growth of cybersecurity into a large industry is the inevitable result of the weaponization of code. For the 20 years from 1994 to 2014, Internet users could enjoy the communication, commerce, and convenience that comes from online life without having to think much about security. With the evolution of more of our life into zeros and ones and the rise of the Internet of Things, cybersecurity needs to be accounted for as a central feature in all the products being developed and commercialized for tomorrow.

Chris Bronk, a cybersecurity expert and a professor of computer and information systems, sees cybersecurity as one of the fastest-growing industries in the world. "It wouldn't shock me if it doubled in size in the next ten years," says Bronk. "It's been doubling." And the companies that really "know what they are doing," like some of the bigger Fortune 500 companies with global concerns that "really know their business priorities," are moving more and more of their IT department over to work on cybersecurity issues, Bronk adds. "Things like managing data centers and email, and providing support to users—that's becoming less labor intensive and the security job is becoming more labor intensive. So I'd say doubling in the next five to ten years is a conservative estimate."

A dozen years ago, the size of the cybersecurity market was just $3.5 billion. Research reports valued the global cybersecurity market at $64 billion in 2011 and $78 billion in 2015; they project it to be at $120 billion by 2017. "Spending, globally, continues to remain elevated as the cyber security market seeks greater definition and maturity; with the rapid pace of the development of solutions in response to the evolving threat landscape, the future horizon for security is ever broadening," one report reads.

I expect the total market size of the cyberindustry to increase even faster, reaching $175 billion by the end of 2017.

Peter Singer, a cybersecurity expert at the New America Foundation and coauthor of a comprehensive book about cyber, *Cybersecurity and Cyberwar: What Everyone Needs to Know*, sees the industry's growth mirroring that of the Internet. "I think it will continue to grow most likely on an exponential curve because it's going to follow the Internet itself. . . . If five billion new people are coming online, five billion new people's security problems are coming online," he explains. "This was a domain, a place that didn't exist when I was born that is now integral to global commerce, integral to global communication, and integral to conflict."

Finnish cybersecurity expert Mikko Hypponen says, "I believe it's going to be as big a shift, for defense industries and for the militaries of the world, as was the technological shift that we've seen since the Second World War." He adds, "If you think about the equipment we were using in World War II and you think about the equipment today, we've seen a massive technological shift. Now we are entering a similar new era where the big shift over the next fifty years, sixty years, will be the development of cyber-ops, completely virtual arms, arms that have nothing you can touch. It's the beginning of the next big shift for the militaries, and it's going to be different because . . . it's much more accessible. You have only less than ten countries in the world which have nuclear arms. Every single country, in theory, can have cyber arms."

With such a big shift under way, Peter Singer cautions that there's a danger of cybersecurity becoming the next military-industrial complex. If handled wrong, the massive growth in the cybersecurity industry could hinge on cybersecurity experts capitalizing on our lack of technical knowledge just as hackers do. "We've got this proto-cybersecurity industrial complex, I call it the cyber-industrial complex, that may parallel the broader defense-industrial complex, who are equally kind of taking advantage of our ignorance and fear," he says. He cites as evidence the hike in lobbying dollars going toward these issues. A decade ago, only four companies were lobbying Congress on cybersecurity issues. By 2013, that figure had jumped to 1,500, says Singer. "There is a gravy train going of people making money off this, and it sometimes aligns with government bureaucracies that are interested in heightening the perceptions of the threat to drive budget dollars for them."

The development of a cyber-industrial complex that mimics the military-industrial complex in terms of its dominance would reach into the computers, tablets, and mobile phones of every Internet user. Singer is justified in being vigilant, but I think the development of a cyber-industrial complex is unlikely for several reasons. First, the development of multibillion-dollar weapon systems characteristic of the military-industrial complex does not match up well with the nature of the weapons and conflict in cybersecurity. The ability to be fast, dexterous, and effective will matter more than which member of Congress you have a relationship with when it comes to winning a contract. There will be a lot of money to be made from government growth in cybersecurity, but the companies that government will want to work with will be those that are innovating quickly—which puts the lumbering bureaucracies of the military-industrial giants at a disadvantage.

Bronk has a hybrid view, expecting dexterous start-ups to be rolled up into large companies much as the defense giants of the military-industrial complex were, but with a modern, Silicon Valley twist. "What I've noticed," he says, "is that good cybersecurity comes from smart researchers, and smart researchers tend to agglomerate with one another in small groups and start-ups." Eventually, Bronk thinks, a big company or a big defense firm bulking up their respective cyberdefenses could buy up or invest in these smart researchers.

"Basically cybersecurity is going to be very similar to Silicon Valley start-up and acquisition patterns," says Bronk. When a Silicon Valley company wants innovation, they either do the work in house or contract it out. "But getting companies to think differently about what they do and radically undercut their current way of business in some area and do something radically different to make much more money, that isn't really in the culture of a lot of companies," he adds.

Bronk recalls talking to a Cisco executive several years ago about what companies do when they want to develop something unorthodox, noting that the Defense Department faces similar innovation issues. The executive told him that the way his company gets around this is to look for an employee at Cisco with a good idea; then it puts that person on a leave of absence, hooks him or her up with a venture capital firm in Silicon Valley, and gives the person a year or two to work on the great idea. "And then if you build it and it works, Cisco gets the first crack at buying it. That's how Silicon Valley, in my mind, largely works," says Bronk. "A little company builds up, innovates, builds a model of business some way or innovates a new product, and then some big guy either pumps a lot of venture capital into it and they go public and become a megacompany like Facebook or Google or Apple or whoever, or they're acquired by a megacompany and they become a division of Facebook or Google."

Security firms are no different, says Bronk. "They innovate, they come up with a cool device, and then somebody like Hewlett Packard comes along and says, you know, 'We really need that in our suite of network management tools.' Bam! Now they're a part of whatever product Hewlett Packard is shipping in network management, and they're the security component."

I agree with Bronk's description of how innovation and the accompanying mergers and acquisitions activity work in Silicon Valley. My expectation, though, is that there will be megacompanies (to use Bronk's word) that break through from start-up ranks rather than military-industrial giants that start big and stay big, feasting off the money to be made in big contracts for cybersecurity.

However the growth of the cybersecurity industry takes shape, one point that I have never heard anyone even try to rebut is that the industry is going to get very big very fast. If any college student asked me what career would most assure 50 years of steady, well-paying employment, I would respond, "Cybersecurity." The growth is steep, the need will be sustained, and this ever-growing need currently comes up against a major talent shortage. The qualified job candidates are too few.

The Bureau of Labor Statistics, hardly prone to hyperbole, reports that there will be "a huge jump" in demand for people with information security skills. Echoing a point made by Jim Gosler, the head of a very successful multibillion-dollar hedge fund based in New York that invests in cyber told me, "There's a small group of highly talented people who really understand this stuff to the point where they can actually design hardware, software solutions to actually address them." He explains that the unique thing about cybersecurity right now is that it is not just an issue for one industry or one vendor to tackle. It is an issue that any connected company or individual will have to face at one point or another: "The stakes are big and they're getting bigger for everyone . . . so it's a bigger problem, bigger opportunity, depending on your vantage point."

One vantage point that cannot be forgotten is that of citizens and small businesses that cannot pay for the type of expensive cybersecurity protection that governments and major corporations can. The development of the cybersecurity industry in response to the weaponization of code has been just that—the development of an *industry*. Security, however, is supposed to be a public good administered by government, not a private good purchased in the marketplace. For all the attention newly given to protecting our critical infrastructure like GPS and our large enterprises like banks, there is a huge gap that the market will not solve by itself: everyday citizens and small businesses. Government has a responsibility to protect its people, not just its big businesses and infrastructure.

Government can and should work extensively with the private sector to make sure that the brightest minds are working to develop cyberdefenses, but there is an as yet unmet obligation by government to define its responsibilities to its citizens in this newest domain of conflict. The way the market is constituted today is analogous to a company developing an antiaircraft gun at a time when air bombing is rampant—yet only selling the antiaircraft gun to buyers in the marketplace rather than using it to defend the wider civilian population. This will require some changes in the form of more efforts like Cloud-Flare's Project Galileo or by government-led initiatives to guarantee a minimum level of cybersecurity to all its citizens. We all want the liberty that comes with a vibrant online life, but liberty without security is fragile, and security without liberty is oppressive. The years ahead will force us to balance these two as we have not had to before.

FIVE

DATA: THE RAW MATERIAL OF THE INFORMATION AGE

Land was the raw material of the agricultural age.
Iron was the raw material of the industrial age. Data
is the raw material of the information age.

At the house that I grew up in, the woods ran right up to our back door. Many summer mornings, I would run out the door and into the woods with my buddies. We would play for hours, roaming over a few square miles of forest until we got hungry. Then we'd make our way home to gobble up a lunch of macaroni and cheese and head back out into the woods until dinnertime. Our parents had a general sense of where we were, but they didn't care exactly where. We were unmonitored. We were untrackable. We were unreachable. There were no adults anywhere. Just kids and animals. Our parents knew we'd come home when we got hungry.

That was the normal state of affairs for children of my generation. Suburban kids would hop on their bikes; city kids would hit the playgrounds and the subways. Today every kid has a cell phone, including my 13-year-old son. When today's kids leave the house, they are in constant contact with parents and friends by phone and text. They're emitting GPS signals. They're leaving digital footprints on social media. They are little beacons of data production and consumption. If any of my three kids went off the grid in the same way that I used to, my wife and I would be frantic that something had gone wrong.

We have adjusted to a reality where everyone is reachable at all times, even our children, and we expect and demand to be plugged in at all times. I don't know if this is a good thing or a bad thing— probably a bit of both. Either way, we are at a pretty remarkable inflection point in history. There is already a fundamental generational difference between the always-off childhood of my generation and every one that came before, and the always-on childhood of my kids and all who will come after.

The first time a child is handed a phone or plays his first video game, he begins building a stack of personal data that will grow throughout his lifetime, a stack that can be constantly collated, correlated, codified, and sold. I didn't send or receive a single email or text message when I was in college just over 20 years ago. I didn't post anything on social media. I didn't own a cell phone. Even so, I'm now thoroughly catalogued and monetized like the majority of Americans. Private companies now collect and sell as many as 75,000 individual data points about the average American consumer. And that number is tiny compared with what's to come.

The explosion in data creation is a very recent occurrence, and from its inception, data storage has grown exponentially. For millennia, record keeping meant clay tablets, papyrus scrolls, or parchment and vellum made from animal skin. The first modern paper, made from wood or grass pulp, was a big advance; but the first major milestone in the mass production of data came with the invention of the printing press. In the first 50 years after the first printing press appeared, 8 million books were printed—more than all the books produced by European scribes in the prior millennium.

With the successive inventions of telegraph, telephone, radio, television, and computers, the amount of data in the world grew rapidly during the 20th century. By 1996, there was so much data and computing had gotten sufficiently inexpensive that digital storage became more cost-effective than paper systems for the first time.

As recently as 2000, only 25 percent of data was stored in digital form. Less than a decade later, in 2007, that percentage had skyrocketed to 94 percent. And it has continued to rise since.

Digitization dialed up the possibilities for data collection in a remarkable way. Ninety percent of the world's digital data has been generated over the last two years. Every year, the amount of digital data grows by 50 percent. Every minute of the day, 204 million emails are sent, 2.4 million pieces of content are posted on Facebook, 72 hours of video are posted on YouTube, and 216,000 new photos are posted to Instagram. Industrial firms are embedding sensors into their products to better manage their supply chains and logistics. The sum of all this is the creation of 5.6 zettabytes in 2015. A zettabyte is 1 sextillion (10^{21}) bytes, or 1 trillion gigabytes.

Big data is a catchall phrase used to describe how these large amounts of data can now be used to understand, analyze, and forecast trends in real time. The term can be used interchangeably with *big data analytics*, *analytics*, or *deep analytics*. There is a common misconception that the advances made possible by big data are simply a function of the amount of data gathered. In actuality, the growth in the amount of data without the ability to process it is not useful in and of itself. Even in my relatively data-light childhood, my friends and I were still generating plenty of academic data with every test we took and report card we received, but there was no way to connect these data points together and analyze them. The same is true in business; think of all the letters and telegrams that carried business data in the past. They held huge amounts of information, but none of it searchable, none of it usable on a mass scale. The value derived from big data is partially a function of the amount of data created, but as or even more important is our new ability to use that data in real time to make smarter, more efficient decisions. Big data is further aided by new developments in data visualization that allow humans to see and understand patterns that might not be apparent on a spreadsheet full of numbers.

Barack Obama's two presidential campaigns were what first brought big data to life for me in a meaningful way. The story of the campaigns' use of big data is by now well known. In very competitive elections, the Obama campaign used big data to gain insights into how to raise money, where to campaign, and how to advertise, which none of its opponents could rival. From fund-raising to field operations to the analytics in its polling operation, a group of several hundred digital operatives and data scientists crushed their Republican opponents. In 2012, the Obama campaign's voter targeting and turnout programs performed brilliantly, while the Romney campaign's crashed.

Over the course of the 2012 campaign, Obama's 18-person email team tested over 10,000 versions of email messages. In one instance, the campaign ran 18 variations of a single email, all with different subject lines, to determine which would be most effective. The most successful subject line, "I will be outspent," raised $2,673,278. The lowest-performing version of that batch—subject line "The one thing the polls got right . . ."—raised only $403,603.

This was not something that the campaign could just intuit. As a senior member of the email team admitted, "We basically found our guts were worthless." The results of this data-driven fund-raising were impressive. In 2012, the Obama campaign raised $1.123 billion total, with $690 million coming online from 4.4 million donors. The Obama operation was about twice the size and produced outcomes four times as large as Republican Mitt Romney's campaign.

At the head of Obama's campaign was analytics director Dan Wagner, who encapsulates his strategy in one brief anecdote from his childhood: "I was pinning a hoist with my dad one time at my house in Michigan. I think I was building a ramp or something like that.

And my dad looks at me and he says, 'Son, if you have sixty seconds to do something, you have ten seconds to figure out a better way of doing it.' "

Wagner took his father's advice to heart, and he put it to practice in the 2012 campaign: in a nutshell, everything he did was about figuring out how to pin a hoist more efficiently and effectively. Wagner's approach is one that more and more companies are waking up to, especially as data gets cheaper and more flexible. Traditionally, Wagner says, it "isn't the normal psychology of an organization to say, 'If I have this much budget to do something, I'm going to dedicate a certain portion of that budget toward figuring out whether the other portion of that budget is doing a damn thing.' It's, unfortunately, not normal. But I think that's becoming a new normal. You need to reserve a set-aside in your budget to figure out whether the other part of your budget is accomplishing what you want. And it's just an increase in overhead, with the promise of achieving returns on your assets. And you can do that now, because a lot of that stuff is measurable through data that's available and used to not be available."

Beyond newly available data—according to another architect of Obama's data-driven campaigns, Michael Slaby—big data is spurred ahead by new advances in computing: "We've been gathering lots of data for a long time. So what 'big data' really means is the capacity to process lots of information in something that approximates real time so that we can actually do something with it. We can make different decisions based in a strategic way rather than purely a retrospective analysis, after the fact. Typically large data analysis is always only done in retrospection, like a big study or some long-term, longitudinal whatever, rather than something that is a part of an existing strategic process."

Think about the difference between the years-long analysis of the national census and the real-time analysis required if you want to get out the vote in an ongoing national election. Speed makes entirely new projects possible. That's what's remarkable about the emergence of big data to Slaby: "Big data's really just the application of the commodification of computing power combined with the wider availability of cloud computing. We can now chew through enough data fast enough in a way that people can afford . . . and storage got cheap, so we can store lots of data . . . and then we can actually process it fast enough to make use of it."

Increases in data gathering and growth in computing power complement each other. The more data there is, the more investment there is in powerful computers and abundant storage to chew through the data and draw business intelligence from it. The more powerful computers are, the easier it is to gather large amounts of data and produce larger and more in-depth data sets.

Big data is inherently contradictory. It is both intimate and expansive. It examines small facts and aggregates these finite facts into information that can be both comprehensive and personalized. Academics have likened it to both a microscope and telescope—a tool that allows us to both examine smaller details than could previously be observed and to see data at a larger scale, revealing correlations that were previously too distant for us to notice.

The story of big data's real-world impact to this point has been largely about logistics and persuasion. It has been great for supply chains, elections, and advertising because these tend to be fields with lots of small, repeated, and quantifiable actions—hence the "recommendation engines" used by Amazon and Netflix that help make more precise recommendations to customers. But these fields are just the beginning, and by the time my kids enter the workforce, *big data* won't be a buzz phrase any longer. It will have permeated parts of our lives that we do not think of today as being rooted in analytics. It will change what we eat, how we speak, and where we draw the line between our public and private personas.

HOW MANY LANGUAGES DO YOU SPEAK?

One of the things big data is going to do in the next ten years is allow everybody reading this book to be conversant in dozens of foreign languages, eliminating the very concept of a language barrier. It used to be the case when I traveled abroad that I would take a little pocket dictionary that provided translations for commonly used phrases and words. If I wanted to construct a sentence, I would thumb through the dictionary for five minutes to develop a clunky expression with unconjugated verbs and my best approximation of the correct noun. Today I take out my phone and type the phrase into Google Translate, which returns a translation as fast as my Internet connection can provide it, in any of 90 languages. The result is usually good or good enough. When I can't say something, I'll hold up the screen and my non-English-speaking counterpart can read what I am failing to say. Of course my pronunciation is bad to the point of incoherence, I can only do a couple of sentences at a time, and I have no idea what others are saying to me in response. It's basically good enough to ask where the bathroom is and then hope somebody points in the right direction.

Today's machine translation is leaps and bounds faster and more effective than my old dictionary method, but it still falls short in accuracy, functionality, and delivery. In essence, this is little more than a data and computing problem. Professional translators argue that local dialects, inflections, and nuance are too complex for computers to ever account for sufficiently. But they are wrong. Today's translation tools were developed by computing more than a billion translations a day for over 200 million people. With the exponential growth in data, that number will soon signify the number of translations made in an afternoon, then in an hour. Massive amounts of language data will go in and out. As the amount of data that informs translation grows exponentially, the machines will grow exponentially more accurate and be able to parse the smallest detail. Whenever the machine translations get it wrong, users can flag the error—and that data too will be incorporated into future attempts. We just need more data, more computing power, and better software. These will come with the passage of time and will fill in the communication gaps in areas including pronunciation and interpreting a spoken response.

The most interesting innovations in machine translation will come with the human interface. In ten years, a small earpiece will whisper what is being said to you in your native language near simultaneously to the foreign language being spoken. The lag time will be the speed of sound. Undetectable. The voice in your ear won't be a computer voice, à la Siri. Because of advances in bioacoustic engineering measuring the frequency, wavelength, sound intensity, and other properties of the voice, the software in the cloud connected to the earpiece in your ear will re-create the voice of the speaker, but speaking your native language. When you respond, your language will be translated into the language of your counterpart either through his or her own earpiece or amplified by a speaker on your phone, watch, or whatever the personal device of 2025 is.

Today's translation tools also tend to move only between two languages. Try to engage in any sort of machine translation exercise involving three languages, and it is an incoherent mess. In the future, the number of languages being spoken will not matter. You could host a dinner party with eight people at the table speaking eight different languages, and the voice in your ear will always be whispering the one language you want to hear.

Universal machine translation will accelerate globalization on a massive scale. While the current stage of globalization was propelled in no small part by the adoption of English as the lingua franca for business—to the point where there are twice as many nonnative English speakers as native speakers—the next wave of globalization will open up communication even more broadly by removing the need for a shared language. Currently when Korean-speaking businesspeople speak with Mandarin-speaking executives at a conference in Brazil, they converse in English. No longer will there be this need, opening the door for nonelites and a massive number of non-English speakers to the world of global business.

Machine translation will also take markets that are viewed as being difficult to navigate because of language barriers and make them more accessible. I think of Indonesia in particular. There are plenty of English, Mandarin, and French speakers in Jakarta and Bali, but very few of them on most of the other 6,000 inhabited islands. If one does not need to be fluent in Javanese (or any of the 700 other languages spoken in Indonesia) to do business in those other provinces, then they are immediately more accessible and outside capital is in turn more accessible to them.

Just across the Banda and Arafura Seas to the east of Indonesia is the resource-rich country of Papua New Guinea. Papua New Guinea is loaded with mineral deposits, agriculture-friendly land, and water teeming with valuable seafood (including 18 percent of the world's tuna), but its 850 languages scare off most foreign investors. Big data applied to translation will change that. It will take economically isolated parts of the world and help fold them into the global economy.

As with any new technology, the rise of universal machine translation will also have its downsides—and two in particular come to mind. The first is the near-obliteration of a profession. The only professional translators in ten years are going to be the people who work on the translation software. Most machine translation programs (such as Google's) continue to rely heavily on human translations, but once the data sets of translations are large enough, the translators won't be needed. The job of professional translator could go the way of the lamplighter and ice delivery man, or it could become like the coal-mining jobs of today; there's still a need for a smaller number who are supervising a machine rather than extracting the coal from the earth themselves. I imagine a reduced number of professional translators working with the machines to account for the slang and shorthand that always enter into the living system that is language. I remember diplomats at the State Department being appalled at my suggestion that we look at machine-translating Hillary Clinton's speeches as she gave them so they could be heard more broadly and in real time. This will NEVER happen, they exclaimed. The diplomats were right that today's real-time solutions are not good enough and that mistakes might even produce some diplomatic hubbub. But they were wrong that it will never happen. It's just a matter of time.

The second downside is the increased risk of fraud. If my voice can be reconstructed in a way that makes the reconstruction difficult to distinguish from my "real" voice, then it opens up new opportunities for fraud—fraud in dozens of languages, no less. In a world with near-universal translation and communication, an ironic side effect may be that we'll need to be able to look somebody in the eye to believe what he or she is saying.

NINE BILLION PEOPLE WILL NEED TO EAT

The potential of big data to bring Urdu, Greek, and Swahili together into a common conversation could improve the world in remarkable ways. Even more impressive, though, is the role that big data might be able to serve in significantly reducing hunger, probably the longest-running challenge for humanity. The World Food Programme reports that one out of every nine people on earth, 805 million people, does not have enough food to live a healthy, active life. As the population grows as expected to more than 9 billion people over the next 30 years, the amount of food produced will need to increase by 70 percent lest the world grow even hungrier. This comes in the midst of climate change, as temperatures rise and potable water becomes an ever scarcer resource (70 percent of freshwater used globally goes toward agriculture).

The best hope for feeding our more populated world comes from the combination of big data and agriculture—precision agriculture. For thousands of years, farmers have worked using a combination of experience and instinct. For most of human history, the phases of the moon were considered the most important scientific input in farming (because of ancient beliefs about the moon's impact on soil and seed and for the more practical reason of managing time without a clock or calendar). After World War II, a period of scientific and technological innovation spurred what is known as the Green Revolution, which massively increased agricultural production and reduced both hunger and poverty. The Green Revolution introduced new technologies and practices involving hybrid seeds, irrigation, pesticides, and fertilizer. Even since then, though, farmers have tended to work off a fixed schedule for planting, fertilizing, pruning, and harvesting their crops without much regard for changing weather and climate conditions or the changing little details in each field; farming as an extension of the industrial age.

The promise of precision agriculture is that it will gather and evaluate a wealth of real-time data on factors including weather, water and nitrogen levels, air quality, and disease—which are not just specific to each farm or acre but specific to each square inch of that farmland. Sensors will line the field and feed dozens of forms of data to the cloud. That data will be combined with data from GPS and weather models. With this information gathered and evaluated, algorithms can generate a precise set of instructions to the farmer about what to do, when, and where.

When I climbed into a tractor or harvester as a kid, it was a simple, sturdy machine: just a steel frame, big rubber tires, and an engine. The farmer would work the field based on the day and the time, and he'd do it by eyeballing the patch of farmland in front of him. The farm equipment being built for tomorrow looks more like an airplane cockpit than the tractors I remember from childhood. There are graphic interfaces for software programs running on a tablet computer in the farmer's line of sight. The machine moves not based on where the farmer is steering it, but by the instructions coming from software, guiding it by remote control. As the machine works the field, active sensors located beside the headlights feed information about the crop canopy into the system. Navigating itself through the field, the machine is constantly absorbing and applying information—from satellites far above and from the soil below. The instincts are algorithmic. The machine operates with a level of precision beyond even the wildest dreams of farmers from any other point in human history.

Today's versions barely hint at what's possible. Eventually that tractor will be able to sense what each square inch of ground needs and will send out tiny streams of customized fertilizer mixes depending on what that one inch wants. Rather than plastering a field with a fixed amount of phosphorous or nitrogen, it will parse out the amount at the precise level needed.

The early investment to make precision agriculture take root at a global scale is coming from the largest agribusinesses, including Monsanto, DuPont, and John Deere. Monsanto quickly arrived at a strong conviction about the importance of big data and has gone on a buying spree, paying billions of dollars to acquire farm-data analytics companies. It figured out that analytics could increase crop production 30 percent, creating a $20 billion economic impact.

One of the products Monsanto is just beginning to field-test, which gives a sense of what will one day be mainstream, is FieldScripts. FieldScripts takes information about a given farm, including its field boundaries, yield history, and the results of fertility tests, and breaks that farm into a number of small management zones. An algorithm then recommends specific seeds and provides a seeding prescription (when, where, which kind of seed, and how much) that is delivered to the farmer over an iPad app called Field View. The farmer transmits data from Field View to a big birdlike piece of farm machinery called a variable-rate planter, with a tractor-like cab pulling two heavy steel wings that distribute seed across an area of farmland up to 30 feet wide. Inside is a monitor that looks as detailed and sophisticated as the dash of a new Boeing 787. Guided by the data, the planter then disperses the customized seeding to each little management zone.

Innovations like FieldScripts are going to make farmers look and work more like office workers. They will spend more of their day focused on tasks like data integration and keeping software up to date and less with their hands in the dirt. In 2014 Monsanto's chief technology officer, Robb Fraley, said, "I could easily see us in the next five or 10 years being an information technology company."

While huge agribusinesses are smart to recognize the market opportunity of bringing big data to the farm, it looks doubtful that they are going to dominate this new sector as they do the sale of tractors and fertilizers. I think back to the early days of personal computers when large established companies like IBM and Compaq with networks of business customers were able to make the early investments that cemented their market leadership. Despite their dominant position, they eventually lost their market share on the hardware side to upstarts like Apple and Dell and on the software side to companies like Microsoft that had the benefit of being born specifically for the personal computer market. Monsanto, DuPont, and John Deere will have to continue to acquire the most promising start-ups, the precision-agriculture natives, if they are going to stay ahead in what will be a fast-changing field.

I also think that small farmers are likely to derive as much benefit from precision agriculture as larger farmers with thousands of acres. Precision agriculture is not based on huge enterprise software systems that take up half the barn. That expensive software is in the cloud and accessible on cheap devices like smartphones and the tablets I saw in the tractor's "cockpit." The costs on the hardware side, including the sensors, will continue to decline, and the real costs will come from subscriptions to the software as a service—the precision agriculture content. This is the business model the big agribusinesses are already pushing, and it will spread from the highest-tech farmers working huge fields to the small family farm.

It will still take years for this kind of technology to mainstream in wealthy parts of the world, but it will happen. Not long thereafter, it will come to the developing and frontier parts of the globe. The pattern will look similar to robotics, with what are initially high up-front capital expenses that create offsetting savings in operating expenses. As the cost of equipment goes down, it will be more accessible to farmers in the developing world. This is where its impact may be greatest. India comes to mind. No other country has suffered more from a lack of farm modernization than India. Farming is so difficult in India—because of high costs, scarce water, and other factors—that an estimated 300,000 farmers in India have committed suicide in the past 20 years. What's more, 25 percent of the world's hungry, an estimated 190 million people, live in India, and hunger is the country's number one cause of death. As the population continues to grow rapidly, the task of producing enough rice, wheat, and other staples grows more difficult.

Precision agriculture will not end hunger in India or turn its subsistence-level farmers into serious agribusinesses, but in an environment of scarcity, it can take those scarce resources, be they seed, fertilizer, or water, and get the most out of them. India does not have a national network of agronomists to provide expertise and resources to its country's farmers as China, the Americas, and Europe do. The budgetary resources in India are spread too thin. When precision agriculture mainstreams, it could play the role of these national networks of agronomists. The best hope for India is that precision agriculture provides a leapfrog opportunity, helping its subsistence-level farmers achieve a level of performance that is impossible for them today. It represents the best hope for feeding the hundreds of millions of people in India who don't have enough to eat today. Precision agriculture will take farming from being an industrial age industry to a digital age industry.

Precision agriculture also offers the promise of a major reduction in pollution. My adopted home state of Maryland provides an excellent example. I live on the Chesapeake Bay, home to beautiful waterways and delicious blue crabs. But the population of the bay's crabs and other animal life has plummeted in recent decades because nitrogen from excess fertilizer has fouled the water. Nitrogen from fertilizer creates dead zones; there is one in the Gulf of Mexico that's the size of New Jersey. And as the fertilizer kills off my home's aquaculture, it also contributes to climate change. Fertilizer produces nitrous oxide, a nasty greenhouse gas that is right up there with carbon dioxide and methane in terms of its adverse impact on the climate. Excessive fertilizer in farm fields pumps nitrous oxide into the atmosphere. Yet none of the climate change agreements being negotiated account for farm-based nitrous oxide in any meaningful way because of the worry that regulating fertilizers would reduce the food supply and contribute to hunger.

Precision agriculture offers an alternate solution. Instead of blanketing a field with a fixed amount of fertilizer, pesticide, and herbicide, new data will allow us to significantly reduce the amount of chemicals we put on farm fields, which will in turn reduce the amount that ends up in water, air, and food. By using local sensor inputs to determine just the right amount of water and fertilizer to use, precision agriculture holds the promise of growing more food while polluting less, all with the help of big data.

FINTECH: THE FINANCIAL DATA SYSTEM

Wall Street has taken advantage of big data as much as any industry. Of the roughly 7 billion shares that are traded in US equity markets every day, two-thirds are traded by preprogrammed computer algorithms that crunch data about share prices, timing, and quantity in order to maximize gains and minimize risk. This is called black-box or algorithmic trading and is now the norm in finance.

The next impact of big data in the finance world will be in retail banking, the area where average people are the customers, as opposed to investment banks or commercial banks that focus on serving corporations. The application of big data to enhance operations and product development in retail banking is known as "fintech."

The technology that undergirds the banking system's current infrastructure is obsolete. As Shakil Khan, the creator of the Bitcoin news service CoinDesk, describes, the systems behind basic banking functions like processing loans and monitoring accounts are "very, very old in today's day and age. Where technology companies are innovating and bringing out new products every six months, banks are running on systems that were built in the 1980s and 1990s."

Fintech seeks to change that. In 2008, financial technology firms raised approximately $930 million in investments globally. In 2013, these firms collectively raised about $3 billion and that amount is expected to reach $8 billion by 2018. Fintech start-ups are also forcing large institutional banks like HSBC, UBS, Barclays, and JPMorgan Chase to invest in their own technology overhauls. The clunky technology inside many big banks was on full display during the financial crisis when bank CEOs and regulators were reduced to guesswork about their own loan portfolios. When asked why they could not better understand the mortgages they were carrying, the bank officials always pointed to the dozens of legacy technology systems that were not interoperable.

"We owe it to ourselves, our customers, our clients and our shareholders to cast a wide net and find new ways to solve existing and new challenges," said David Reilly, a technology infrastructure executive at Bank of America. "Engaging with the start-up and venture capital community forces us to think about innovation in a different way, more revolution than evolution."

One representative of the fintech community is 29-year-old Zac Townsend. Zac cofounded Standard Treasury, a start-up company recently bought by Silicon Valley Bank that was established to figure out how technology can help banks better interact with their customers. Over breakfast in downtown San Francisco, Zac explained what brought him to fintech. Born and raised in New Jersey, he started a career in government innovation, working for both Mayor Michael Bloomberg of New York City and Mayor Cory Booker of Newark. But rather than move to Washington, D.C., when Booker was elected to the US Senate, Zac decided to head to Silicon Valley. As he explained, "I thought to truly understand innovation in government, I had to actually understand innovation." Within a year of his move to the Bay Area, Zac had lost nearly 50 pounds on a low-carb diet and raised a couple of million dollars in venture capital. And he was at the center of an even bigger transformation.

According to Zac, big data is on its way toward radically changing consumer banking. "What is a bank?" Zac asks. "A bank is a giant ledger that records how much money belongs to people and how much money people owe them. At heart, that's a data problem. I think that banks—and the ecosystem around banks—are just starting to realize that they're digital companies. They are data companies. They are technology companies. That really hadn't occurred to them in the past even though so much of what they do is data based. They use Bloomberg terminals, they price risk, they analyze markets. There is just so much data. But the reality is that they are just starting to see data as their core business."

Zac argues that the financial industry has been hampered by antiquated data systems, which have made them opaque to customers. "Banks and hospitals are the only places in the world where you walk into them and you don't know what they sell. There is no list of products. There is no list of prices. So, if you are an underbanked person in Oakland and you walk into a local branch of, say, Wells Fargo, there's columns, there's carpeting, there's desks. But you have no idea what to do. It is no surprise to me that you would be intimidated and have trouble opening a bank account. On the other side of the world there may not even be established banks. People are still worried about how they store value or pay people."

Zac's start-up, Standard Treasury, began as an effort to design a particular type of software program, application programming interfaces (APIs), for banks that allow them and their customers to better access, use, and visualize data. This is just one step toward the bank of the future. Zac thinks the banks that emerge in the 21st century will be those with their origins and capabilities in managing big data. "What banks do—storing value, moving value, and pricing risk—those are all the functions of a data company. Google could, if they wanted to, do those three things better than most banks," he says.

And that raises an interesting question. Why *hasn't* Google—or some other analytics firm—created a digitally native bank?

It's a question Zac is intimately familiar with. "The problems are primarily regulatory," he explains. "It is very hard to buy a bank. It is very hard to capitalize it. Banks are very restricted in their operations. Regulators are very scared of innovation. They are very scared of a bank that does not make money in the way that other banks make money. The reason they're uncomfortable is the reason we all should be uncomfortable: banks aren't really meant to make money. They're meant to serve as rock-solid repositories for other people's money. I have talked to people at the FDIC. I have talked to people at the Office of Control of Currency. I have talked to lawyers and regulatory consultants. The question that you come down to is not, 'Can you buy a bank?' I could probably raise a couple million dollars and buy a small bank. The question is, 'Could you actually operate a bank like a technology company?' The answer is probably no. And I'll give you the simplest reason, which is what every regulator has told me: they are uncomfortable with a bank that grows more than 20 percent year over year. The number one historical indicator of a bank that is about to go belly-up is one that has a lot of growth. But a good Silicon Valley start-up grows 20 percent month over month."

Regardless, Zac made an effort. The room for improvement was too enticing for him to give up. Faced with regulatory hurdles to make a go of it as a bank in the United States, Zac's company applied to become an independent bank in the United Kingdom, describing itself as a "technology-first wholesale bank." He then sold Standard Treasury to Silicon Valley Bank, adding traditional banking infrastructure but doing so with a bank that thinks "digital first." Asked about competing against the centuries-old UK banks, Zac told me, "We are competing with Lloyds, HSBC, RBS, and ultimately I'd hope JPMorgan Chase and Wells Fargo here in the United States. They are our competition, and like anyone from New Jersey, I intend to destroy them." In Zac's scenario, technology-first banks could provide a more people-centered service that tears down the old big banks the way that increasingly convenient options like digital photography destroyed Kodak and that Google and Wikipedia nearly eliminated leather-bound encyclopedias.

Zac firmly believes that the older legacy banks would have avoided the catastrophe of the banking crisis if they had used big data more effectively. He explained his view to me: "Basically they're just bad at basic questions of data management and trust. Who owns this mortgage? Has the lendee been paying regularly? What is my total exposure to this single customer? For example, right now globally regulators are beating down the doors of big banks to create 'single customer views,' which basically means, 'Can JPMorgan Chase say the total liabilities (deposits) and assets (loans) to Exxon?' They currently have a lot of trouble doing that because they're working across so many legacy systems that they have trouble rolling up their analytics to one central view. No wonder they've had some troubles!"

"They aren't managing their data well," Zac adds, "and that should scare people. A bank is a little more than data. They are big data companies! You can call them fancy things like securitized subprime mortgages or credit default swaps or whatever, but a bank is basically a giant ledger of contracts that have future positive and negative cash flows. A bank's entire income is based on how the present value of those cash flows changes moment to moment yet they can't get the basic accounting of who owes what to whom correctly."

Zac is not alone in making this critique. A generation of young entrepreneurs like Zac is trying to change how the financial system works by smartening it up through big data. Zac's work is not far off from that of Jack Dorsey and his colleagues at Square, who now believe that they are better positioned to judge loan candidates than a traditional bank is because they have a real-time look into all an applicant's transactions.

Square now has a program, Square Capital, that goes beyond facilitating transactions and gives its users access to capital so that they can grow as their demand rises. Jack explains, "Now they can actually open up a register, the thing that runs their business, and they can hit a button and they can get an instant loan for $10,000. And then they pay back that loan—with a very, very low interest rate—by swiping their customers' cards." In other words, the loan is easy to take out and almost automatic to pay back, through small payments tied to each sale. Jack says, "The reason we can do this is because we know their business. We see all their cash transactions. We see their card transactions. We see their check transactions. So we know if their trend is going up. We know their area. And we can make really, really smart decisions around how much we can give them, if this is going to be something that they'll pay back in a reasonable time frame. . . . So it's a very, very simple way of using the heart and the heartbeat of their business, which is the register."

In a coded-money economy, a lender knows a merchant's true value because it has real-time access to its books. Instead of checking a credit score, the data for every dollar going in and out of the store is immediately available. That lender knows a merchant's value without ever having to physically open up his or her books for inspection. Square board member and investor Roelof Botha says, "We have a huge competitive advantage. No one else has the kind of visibility we have into how these businesses that use Square are doing." The visibility Botha describes comes from the more than $30 billion in payments it is now processing yearly, which has enabled Square Capital to lend more than $100 million to some 20,000 merchants within its first year of operation. This provides Square with a business model that is much more remunerative than one that comes from the narrowing margins in credit card transactions.

Zac Townsend and Jack Dorsey are betting that the big data revolution will open up opportunities for breakthroughs in finance far beyond payments. Zac says, "I think we are entering an era in Silicon Valley where people are attacking big, meaty problems. This is in no small part because of big data. It may seem like we are working on seemingly boring things—like improving financial technology—but it turns out that tens of millions of people's lives are worse off because they haven't had access to effective or efficient services. With better analytics, we can help change that."

ALL-SEEING STONES

A prime example of the "big, meaty problems" that Silicon Valley is now taking on—which at first look "seemingly boring"—can be found in the work of a secretive tech company that provides a glimpse of some of the scarier applications of big data.

While I was at the State Department, a captain in the US Marine Corps was detailed to my office. He had extensive battlefield experience as a sniper and was assigned to my office to learn how technology might be applied to help Marines on the battlefield in Afghanistan. His task was simple: identify ways in which the information environment might be enhanced for Marines on the ground in isolated parts of Afghanistan, so that the Marines kill more Taliban fighters and the Taliban fighters kill fewer Marines.

The captain and his colleagues got behind a technology from Palantir, a Palo Alto–based company named after the Palantiri all-seeing stones in *Lord of the Rings*. The company is run by Alex Karp, an eccentric Stanford social theory PhD whose hobbies include solving Rubik's Cubes and qi gong meditation. Karp was a student under Jürgen Habermas, the German philosopher and sociologist famous for his notion of the public sphere and its importance as a free discussion forum where public opinion is formed. From 2005 to 2008 the CIA was Palantir's sole customer. Since 2010, Palantir has also designed software systems for the NSA, the FBI, and the US military.

Palantir specializes in data management, transforming massive and often messy data into visualized maps and charts. For Afghanistan and Iraq, Palantir designed a technology with multidimensional maps that took in information about the timing, severity, and targeting of attacks in order to plot out patterns of risk that could be read and understood in real time. Palantir's services have since extended well beyond predicting insurgent attacks. Its technology has been used to analyze roadside bombs, sniff out members of drug cartels, and track cyberfraud. Karp now travels with a 270-pound bodyguard even in the placid confines of Silicon Valley.

Palantir has also moved into the corporate world, helping companies in financial services, legal research, mortgage fraud, and cybercrime through its analytics program. It describes its core disciplines as "data modeling, data summarization and data visualization." All three of these are crucial to navigating complex environments quickly, and they represent what the field of big data is capable of. It is not just being able to scour a large database; once data has been analyzed, it needs to be summarized in an easy-to-understand way and presented visually so that humans can then apply their own expertise and make their own judgments. A Marine in Afghanistan is not going to be able to look at a spreadsheet of numbers and know which street in which village is the likely site of an ambush. The work of big data is to chew through that spreadsheet, summarize the information, and display the findings in a map, a graph, or some other picture that the Marine will grasp immediately. Finding patterns that we would otherwise overlook and displaying them in ways that we can't miss is Palantir's goal. An investor prospectus for Palantir revealed that its software was used to scour 40 years' worth of records to help convict Bernie Madoff of securities fraud. While it is not forced to disclose its revenues, the private sector now reportedly represents a majority of the firm's revenue.

Alex Karp regards the work of Palantir as sanctified. When

recruiting, he says, "We tell people you can help save the world." To some extent, that's probably true, but I don't believe that these kinds of capabilities will stay bottled up and only be put to work in the common interest. Perhaps we can trust Palantir not to engage in dodgy projects, but it won't have a monopoly on this kind of intelligence capability. Eventually some Palantir-like company will emerge and focus its mapping and targeting capabilities for malignant purposes. Instead of being used to target fraud, a Palantir-like technology could be used to identify people who would be prime victims of fraud. These technologies are value neutral until a human directs them; they take on the values and intentions of their human masters.

We don't think twice about applying all-seeing analytical programs in combat theaters to help protect Marines, but what about when the all-seeing technologies are looking at us? Former CIA cyber guru Jim Gosler likens big data to a game, saying that "by playing the game that we play because of the benefit of it, we're exposing ourselves to advanced analytics on top of big data to know more about ourselves and certainly more than we'd want other companies to know or maybe the federal government to know about us."

The "game" Gosler refers to involves both the data that we freely give up as well as our personal data that is recorded by companies and governments. Often we give up our data in exchange for free or convenient services, coupled with a vague promise of security—but in recent years, it's become clear that once our data is handed over, it can often be used in secretive or questionable ways.

EVERYBODY WILL HAVE A SCANDAL

Gosler's observation is all the more troubling because digital data is practically indelible. Many of us are learning the hard way that once data is produced, it rarely fades away.

My friend and former State Department colleague Jared Cohen now runs Google Ideas, a think tank started by Google in 2010. He recently became a parent for the first time and is especially concerned about the privacy of children in an age of data permanence. "This is the thing that is most scary to parents," he says. "Whether you are in Saudi Arabia or the United States, kids are coming online at a younger age and they are coming online faster than any other time in history. They are saying and doing things online that far outpace their physical maturity. If a nine-year-old starts saying a bunch of stupid things online, it lives in data permanence the rest of his life. Or if a girl in Saudi Arabia is ten years old and chatting with another ten-year-old boy and maybe she says something inappropriate, not really knowing what it means, that may follow her around like a digital scarlet letter for the rest of her life. And when she is twenty-two and it is taken out of context, the consequences can be very real." This is the kind of worry that my buddies and I never had when we were running around the woods of West Virginia.

Jared points out that the "data talk" will now be a mandatory part of growing up and that it should come even before the proverbial birds and bees talk: "Parents are going to have to talk to their kids about data permanence, online privacy, and security years before they have the first conversation about sex and the birds and the bees because it will actually be relevant years before. Every parent is going to have to figure out how to have what will be, in some respects, the very first serious conversation with their kids—about how the things they do today can impact them tomorrow."

Jared's concern also extends more broadly, into the classroom: "The way our educational system deals with socializing kids has to change," he says. "When I was in elementary school, I remember health class. All they did every day was scare you from using drugs, and as you got older scare you from having unprotected sex. There's a culture ingrained in educational systems to scare you from doing bad things. You should also have the equivalent of health class for helping people understand the risks associated with data permanence so they can make smart decisions themselves."

The dangers out there aren't always obvious, which makes data permanence all the more tricky to navigate. Take the mobile app Good2Go, which markets itself as a "consent app." On its home page, a young man and woman stand in shadow. They're looking down at a phone in his hand. The text reads, "When girl meets boy and sparks fly, and you need an answer to the question: Are We Good 2 Go? The Educational App for sexual consent." The idea behind the app—encouraging men and women to get "affirmative consent" before having sex—is a worthy one. But here's the problem: the app logs the name and phone numbers of its users as well as the level of sobriety and exact time at the moment of "consent." This creates a permanent record of who you are having sex with, at what time, and whether you were sober, tipsy, or drunk. Does Good2Go have the legal right to sell that information to marketers? Yes. The app's privacy policy does not appear on its website, but once you do find it, it reads that the company "may not be able to control how your personal information is treated, transferred or used," even if the app itself disappears.

It is not just your past emails or love life that can come back to haunt you. It can also be the math class you failed, the fight you got into in school, or your inability to make friends as a toddler.

In the United States, many parents are horrified by a Department of Education survey that lists hundreds of intimate questions teachers are encouraged to answer about students. According to the Department of Education, this data helps provide individually structured help to children and guide an overall improvement in education— exactly the type of goals that big data proponents have touted in education. Yet parent groups protest what they see as an intrusion of their children's privacy.

A separate but related concern applies to companies that exhibit a predatory approach in how they acquire and sell data about children. A $100 million database called inBloom, which shared confidential student records with private companies, was shut down only after vocal criticism from parents.

Every few weeks, a new example arises that illuminates the problems accompanying the broader commercialization of our personal data. A 2013 Senate Commerce Committee report described a company that sold lists of families with specific illnesses, including AIDS and gonorrhea. In another example, Medbase200, an Illinois-based company that sells marketing information to pharmaceutical companies, went as far as providing lists of rape victims for $79 per 1,000 names. The lists also included domestic violence victims, HIV/AIDS patients, and "peer pressure sufferers"—and they were taken down only after a *Wall Street Journal* reporter started looking into it.

The rise of big data has reawakened the world to privacy as a public-policy issue. It is difficult to bring big data technologies and the value of privacy into alignment. The difficulty is brought about by both surveillance and sousveillance. Government intelligence and law enforcement agencies vacuum up a huge amount of communications data from above through surveillance. Less discussed, but an even bigger problem for people who are not targets of terrorism or law enforcement investigations, are cell phone cameras and wearable technologies capturing what we do and say "from below"—sousveillance. Our secrets and private lives are just as likely to be shared with the world by somebody self-publishing from their mobile phone as they are to be gathered up and released from above. It is a problem that comes from both government and industry, and also from individuals who now own what were considered military-grade technologies 15 years ago.

In response to worries over how technology is eroding privacy, many European governments have established strong privacy regulations. The problem they encounter when trying to enact these regulations is twofold. First, most of these big data technologies do not collect information and organize its collection or distribution by country. If the app or other big data program is centrally located and stores the data in a pro-business, permissive environment like the United States, the working assumption of many companies is that they only have to operate within the legal jurisdiction of the United States. Second, when countries do try to disallow their companies from building products that impinge on privacy regulations, they are in effect dialing down their ability to compete in one of the fastest-growing segments of the global economy. Restricting access to data in tomorrow's economy is akin to regulating land use during the agricultural age or regulating what factory owners could build during industrialization. These countries find themselves in a double bind: in order for regulation to serve the public interest, it must be enough to protect individual and community rights but not so much as to eliminate investment and economic growth.

Whether or not we may *want* to respect a stronger version of privacy, it's possible that we're now unable to turn back and actually reach that notion of privacy. Margo Seltzer, a professor of computer science at Harvard University, argued at the 2015 World Economic Forum in Davos that, "Privacy as we knew it in the past is no longer feasible. . . . How we conventionally think of privacy is dead." With the proliferation of sensors, devices, and networks sucking up data everywhere, we may be past the point of being able to halt data collection in any meaningful way. Instead, we may have to concentrate on retention and proper use, that is, clearly setting out how long data can be retained and governing how it can be used, whether it can be sold, and the kind of consent required from the person providing the data.

While Seltzer makes the case that virtually every bit of our personal information is now available to those who want it, I do think there are parts of our lives that remain private and that we must fight to keep private. And I think the best way to do that is by focusing on defining rules for data retention and proper use. Most of our health information remains private, and the need for privacy will grow with the rise of genomics. John Quackenbush, a professor of computational biology and bioinformatics at Harvard, explained that "as soon as you touch genomic data, that information is fundamentally identifiable. I can erase your address and Social Security number and every other identifier, but I can't anonymize your genome without wiping out the information that I need to analyze."

The danger of genomic information being widely available is difficult to overstate. All of the most intimate details of who and what we are genetically could be used by governments or corporations for reasons going beyond trying to develop precision medicines. If the trade-off for being able to benefit from lifesaving therapies because of genomics is handing over our most intimate data, then what are needed are strong rules regarding how it will be retained and used.

If the world ten years from now is a world without privacy as we know it today, then norms will shift. In a world without privacy, everybody will have a scandal. And in that world, the very idea of what is scandalous behavior has to change. I think back to the 1992 presidential election, when the question of whether Bill Clinton inhaled after puffing on a marijuana joint became a major campaign issue. Fast-forward to 2008, and Barack Obama's admitted past use of marijuana and cocaine was essentially a nonissue. Norms had shifted during those 16 years.

Over the next 15 years, with more of our lives captured by big data technologies, norms will shift even further. What constitutes scandalous behavior today won't be as novel or newsworthy. We will increasingly have to accept the fallibility that comes with being human, because each of us will have our errors and indiscretions preserved in indelible data. Even with these shifting norms, we'll need to still try to keep essential information like our genetic makeup from becoming public. As big data erodes privacy, there are some things worth fighting to keep private.

OUR QUANTIFIED SELVES

Privacy is just the first in a series of concerns that big data will raise as it inserts itself more inseparably into our lives. There are abundant other reasonable objections to the dangers of a newly quantified self and society.

Philosophically, there has been a longtime fear with the rise of robotics and automation that machines will become more human—potentially supplanting "us" by taking our jobs or by literally taking over.

In the big data world, the new fear is that humans will become more like machines. I think back to the Obama campaign official who said, "We basically found our guts were worthless." We may live more efficient lives as instincts are replaced by algorithms, but it is reasonable to fear that some of our most human qualities—love, spontaneity, autonomy—may be changed for the worse by our living more algorithmic lives.

As it becomes more ubiquitous, *big data* will fade from use as a buzz phrase. As it reaches into more and different aspects of our everyday lives, the combination of big data and behavioral science will subtly change our routines and expectations through a series of digital nudges that guide our choices through the day.

I remember standing in front of my closet when I was in college wondering what I should wear on a date. In the future, it's not inconceivable that a computer program could scan your closet, then query the profile of the person you are going on a date with, and then make recommendations about the clothing in your closet that will be most appealing. It is also quite likely to try to sell you something that is *not* in your closet that has an even higher probability of producing a positive response. Imagine if that algorithm queried data sets purchased from Good2Go and determined where you should go based on where past "acts of consent" had taken place. The person you are going on a date with might have been matched to you through a dating site's algorithm, or maybe it's someone you met at work, where you secured employment through LinkedIn connections.

Serendipity fades with everything we hand over to algorithms. Most of these algorithms are noiseless. They gently guide us in our choices. But we don't know why we are being guided in certain directions or how these algorithms work. And because they constitute the value of a company's intellectual property, there is an incentive to keep them opaque to us.

Having an algorithm select an outfit for a date might seem far-fetched, but let me ask: In the example above, did the outfit algorithm seem more or less far-fetched than the idea of relying on a matchmaking site's algorithm to find a date? Such a big element of our lives—who we date and fall in love with—certainly seems like it should involve more human choice and less computer algorithm. Yet we're already ceding it to algorithm, to the point where an estimated one-third of all marriages in the United States begin with online dating.

Critics like writer Leon Wieseltier warn that "the religion of information is another superstition, another distorting totalism, another counterfeit deliverance. In some ways the technology is transforming us into brilliant fools." It's not just choice that we're in danger of giving away; it's often our own creativity and ownership. If someone is using a mobile app, who owns the data that is being produced? Is it the person using the app and producing the data, the creator of the app, the company that made the mobile device, or the Internet service provider transferring the data? It could be any or all of these, depending on the terms of service you agreed to. When you use Good2Go, there are two acts of consent taking place: one is an act of sexual consent, and the second is an act of consent allowing the app's creator to sell that information.

This is an area of conflict that pervades big data in almost every sector, even the field of precision agriculture. Most large agribusinesses require license agreements that allow them to own the farmer's data and use the information in whatever way serves their purposes. With access to farm-specific data, the agribusinesses now have a new level of power over pricing and insight into performance and land value, farm by farm. The agribusinesses can price their seeds and services so they're just expensive enough to get farmers to pay for them but not enough to put them out of business. In the same way that Square can know how credit-worthy a small retailer is by having real-time access to its books, an agribusiness has omniscience about the well-being of the farmer—and it's possible to use that insight to exploit the farmer financially. As a response to this, a group at Purdue University has started the Open Ag Data Alliance, which promises to "operate with a farmer-focused approach through a central guiding principle that each farmer owns data generated or entered by the farmer, their employees or by machines performing activities on their farm."

Who owns the data is as important a question as who owned the land during the agricultural age and who owned the factory during the industrial age. Data is the raw material of the information age.

DUMB DATA

As powerful as big data is, there are some things that it does not do well and for which there is little chance of meaningful improvement in the foreseeable future. I don't see any developments in big data that will change the old truism that machines are adept at things humans find difficult (such as working 24 hours straight or quickly solving a complex math problem) and humans are adept at things that machines find difficult (such as creativity or understanding social and cultural context).

New York Times columnist David Brooks has pointed out that data has failed to analyze the social aspects of interaction or to recognize context: "People are really good at telling stories that weave together multiple causes. Data analysis is pretty bad at narrative and emergent thinking, and it cannot match the explanatory suppleness of even a mediocre novel."

It is also the case that while analyzing ever-larger data sets will produce outcomes like near-perfect machine translation, it will also produce a larger number of spurious correlations. The larger and more expansive the data sets, the more correlations there are, both spurious and legitimate. And most big data programs do a poor job of identifying which correlations are more or less likely to be spurious. The use of big data to draw inferences that should be evaluated and tested is often neglected in favor of using big data to produce real-time transactions—whether that is a stock trade, an adjustment in a supply chain, or a hiring decision. But not all the trends it finds are rooted in reality—or in the variables that they appear to be. And all the predictions made by data analysis should come with what are called error bars, visual representations of how likely a prediction is to be an error rooted in spurious correlation.

When I talk to most CEOs or investors, they either ignore or don't build error bars and they talk about their data-crunching algorithms as if they were created by divine beings. They weren't. They were created by human beings and are error prone. Big data failed to predict the Ebola outbreak in 2014, and then, once it happened, wildly mispredicted its reach. It failed to predict or detect the outbreak early in large part because the data coming out of the areas of West Africa with Ebola was not coming out in languages that were picked up by the monitoring programs. When Harvard University's big data monitoring project, HealthMap, finally did report the finding, it was because it had picked up on a French-language newswire (old media) and reported it after the Guinean government had already alerted the World Health Organization. Once it was clear that Ebola had become an epidemic, a statistical forecast published by the Centers for Disease Control estimated that 1.4 million people could be infected by the end of January 2015 in Liberia and Sierra Leone. The actual number ended up lower than 25,000. Big data can make big mistakes.

What humans do with the inferences big data produces can also be a test of values. When data goes from being unstructured to structured, it takes on the values and prejudices baked into its formulation. For example, in the future, it will be possible for a program to be built for human resources professionals that cross-matches health indicators with employment worthiness. Should predictive analytics be used to

determine whether to hire someone who's likely to get a certain illness but doesn't currently have it? It seems unjustly discriminatory.
But even if you don't openly account for disease risk factors, you can
inadvertently approximate them simply by looking at traditional factors. If the HR professional is in a large organization and evaluating
solely on traditional hiring criteria, such as projected retention and
the output of workers he or she brought in, there will be a statistical
bias against people with a higher predisposition to illness—as well as
for a number of objectionable criteria, including bias against women
in childbearing years.

Correlations made by big data are likely to reinforce negative bias.
Because big data often relies on historical data or at least the status
quo, it can easily reproduce discrimination against disadvantaged ra-
cial and ethnic minorities. The propensity models used in many algorithms can bake in a bias against someone who lived in the zip code of
a low-income neighborhood at any point in his or her life. If an algorithm used by human resources companies queries your social graph
and positively weighs candidates with the most existing connections
to a workforce, it makes it more difficult to break in in the first place.
In effect, these algorithms can hide bias behind a curtain of code.

Big data is, by its nature, soulless and uncreative. It nudges us this
way and that for reasons we are not meant to understand. It strips us
of our privacy and puts our mistakes, secrets, and scandals on public
display. It reinforces stereotypes and historical bias. And it is largely
unregulated because we need it for economic growth and because efforts to try to regulate it have tended not to work; the technologies are
too far-reaching and are not built to recognize the national boundaries
of our world's 196 sovereign nation-states.

Yet would it be best to try to shut down these technologies entirely
if we could? No. Big data simultaneously helps solve global challenges while creating an entirely new set of challenges. It's our best
chance at feeding 9 billion people, and it will help solve the problem
of linguistic division that is so old its explanation dates back to the

Old Testament and the Tower of Babel. Big data technologies will enable us to discover cancerous cells at 1 percent the size of what can be detected using today's technologies, saving tens of millions of lives.

The best approach to big data might be one put forward by the Obama campaign's chief technology officer, Michael Slaby, who said, "There's going to be a constant mix between your qualitative experience and your quantitative experience. And at times, they're going to be at odds with each other, and at times they're going to be in line. And I think it's all about the blend. It's kind of like you have a mixing board, and you have to turn one up sometimes, and turn down the other. And you never want to be just one or the other, because if it's just one, then you lose some of the soul." Slaby has made an impressive career out of developing big data tools, but even he recognizes that these tools work best when governed by human judgment.

The choices we make about how we manage data will be as important as the decisions about managing land during the agricultural age and managing industry during the industrial age. We have a short window of time—just a few years, I think—before a set of norms set in that will be nearly impossible to reverse. Let's hope humans accept the responsibility for making these decisions and don't leave it to the machines.

SIX

THE GEOGRAPHY OF FUTURE MARKETS

World leaders take notice: the 21st century is
a terrible time to be a control freak.

"We want to create our own Silicon Valley." If there's a single sentence I've heard in every country I've been to, it's this one.

Silicon Valley has been home to technology-driven innovation for a long time, but the 20-year period from 1994 to 2014 was something special. People all over the world witnessed a spectacular level of innovation and wealth creation, all emerging from a small 30-mile long, 15-mile-wide strip of Northern California.

Other states and countries have been attempting to build the "next Silicon Valley" for years now. At this point, there's even a formula. As Marc Andreessen writes:

The popular recipe for creating the "next" Silicon Valley goes something like this:

- Build a big, beautiful, fully equipped technology park;
- Mix in R&D labs and university centers;
- Provide incentives to attract scientists, firms and users;
- Interconnect the industry through consortia and specialized suppliers;
- Protect intellectual property and tech transfer; and,
- Establish a favorable business environment and regulations.

It happens all the time all over the world. And it never works.

When I'm asked, "What can we do to create our own Silicon Valley?" my response surprises many people: "You can't," I say. "It's too late. Silicon Valley has a decades-long head start creating the perfect environment for creating Internet businesses. What you *can* do, though, is position your communities to compete and succeed in those areas of innovation that are still to come"—those described in this book.

The development of fields such as genomics, robotics, and cyber will all benefit from the interventions that Andreessen listed. But for cities or countries seeking to create the next hotbed for any of these fields, there are also broader factors to consider. Building an innovation-rich place like Silicon Valley requires specific cultural and labor-market characteristics that can contradict both a society's norms and the more controlling impulses of government leaders. This chapter is about what it will take to compete and succeed in the industries of the future and which of today's societies are positioned for success.

DOMAIN EXPERTISE

With the industries of the future, new avenues of opportunity for countries and people alike will hinge on domain expertise—deep knowledge about a single industry, which tends to concentrate in specific cities or regions. Detroit has domain expertise in cars, Paris has it in fashion, and Silicon Valley has it in Internet-based businesses.

Domain expertise for the industries of the future is still broadly distributed.

To understand domain expertise, consider the following question: Why do a ridiculously high percentage of Internet companies still come out of Silicon Valley when massive investment is being made around the world to compete with it? Many factors are at play, but domain expertise is the most important. For more than 20 years, the world's best computer scientists have overwhelmingly been based in Silicon Valley. They could have been born anywhere, but they came to Silicon Valley for school (Stanford or Berkeley), employment (which creates a self-reinforcing cycle that concentrates talent), and investment (with the Valley offering far and away the most access to early-stage capital in the world). And they came to be included in a culture and community that placed the computer science engineer at the highest level of social status. The Valley came to be not just any old industrial center, but a kind of beacon—a place that promised not just opportunity but a sense of belonging—and that continues to attract wave after wave of ambitious entrepreneurs.

But nothing like that exists yet for the industries of the future, where the most interesting and important innovations are taking place with greater geographic spread than we see with Internet-based innovation. There are early geographic leaders in each of the fields, but it is still far too early to describe any of these as the winners or losers in the competition to be home to the next generation of innovation. And what concentration there is today is not destined to be permanent.

In the current landscape, the most important work in the commercialization of genomics is clustered around universities where much of the original research and development took place. It is in and around Boston because of Harvard and MIT, Baltimore because of Johns Hopkins, and Silicon Valley because of Stanford and the Universities of California in San Francisco and Berkeley. Walking through the offices of these companies, one can't help but notice how diverse the workforces are. European, Asian, African, and South American employees fill these companies and live in Boston, Baltimore, or California because they all studied at American universities. The other major prong of genetics research is in China. Though it does not have a top university program in genetics, China has done an excellent job recruiting its citizens back home after they have studied abroad. As a result, Beijing is quickly becoming a center of domain expertise in genomics.

In cyber, the most interesting companies are often based proximate to government, where domain expertise was developed inside the best law enforcement and intelligence communities, including Washington, D.C., Tel Aviv, London, and Moscow. Europe's first cybersecurity accelerator, CyLon, was cofounded by two top foreign policy aides to British prime ministers. One of the world's largest cybersecurity companies, Kaspersky Lab, is full of former Russian military and intelligence officers. Israel has many of the best cybersecurity firms, founded by people who got their start in cyber in the Israeli Defense Forces, especially Unit 8200, Yehida Shmoneh-Matayim, the intelligence corps focused on signals intelligence.

In robotics, domain expertise and the early commercial leadership is generally concentrated where there is preexisting domain expertise in electronics and advanced manufacturing—in countries like Japan, South Korea, and Germany.

Yet even as the industries of the future offer new opportunities to rising hotbeds of innovation around the world, it's interesting to watch how Silicon Valley's influence lingers and continues to draw start-ups in almost every industry. Consider the example of digital currency and fintech, industries of the future that blend old world and new. New York and London are the world's two dominant centers of domain expertise in global banking today, and both are home to substantial fintech investment. Over the past five years, the United Kingdom and Ireland were home to 52 percent of all the fintech financing in Europe. And New York drew even larger levels of fintech financing than London, with dozens of deals putting hundreds of millions of dollars into the bank accounts of technology companies trying to make the banking sector smarter.

But when Zac Townsend wanted to start a company focused on smartening up the banking industry, he did not start it in London or New York. He started it in California. It mattered less to Zac that expertise in banking specifically was in New York or London than that the expertise in innovation and its supporting culture was in California. He believed that in order to change the banking system, he had to work with it but away from it—an approach that is also reflected in the broader data. Although New York and London are the global centers for banking, they are respectively second and third in fintech financing behind Silicon Valley, which gets about one-third of all the investments that take place in the fintech space.

This raises an interesting question about just how distributed the industries of the future are going to be. When twentysomethings like Zac decide to start a company and determine that in order to do so they need to be in California, it creates a self-perpetuating cycle. More broadly, Zac's decision to base his new data-driven finance company in Silicon Valley reflects a roiling debate over just how domain expertise will develop in the big data industry and what effect this is going to have on the global economy as a whole. With the major impact that big data is having on almost every industry, the way that big data expertise develops has the potential to change the very nature of business. And investors are placing big bets on two very different answers. Will big data serve to centralize businesses, pulling more industries into the gravitational field of Silicon Valley? Or will it allow more businesses to innovate wherever they are, in effect creating more opportunities in more places around the world than has been possible before now?

On one side of the argument is Charlie Songhurst, who sees the Valley as a burgeoning global empire. Recall the example that Songhurst used to describe the impact of Uber on transport, transferring wealth from the owners of cab companies all over the world to Uber shareholders in what he likens to the payment of tribute to an emperor. Because of Silicon Valley, Charlie says, "Global regional inequality is going to be unlike anything we've ever seen except maybe the comparative power of Rome versus the rest of the ancient world."

While I think Charlie overstates things, he makes an argument that's worth examining. His thesis lines up with a number of other thinkers who believe that Silicon Valley's expertise in software and analytics will swallow up entire industries and cause massive centralization. The founders of Uber had no particular expertise in transportation, but that did not matter because of their ability to build a software and analytics platform. The idea that underlies Songhurst's vision is that Silicon Valley companies could eventually run everything in which software and big data are useful—which is basically every industry on the planet.

So what's going to happen in this new data-driven empire, according to Charlie?

"It's a very simple equation," he says. "Countries with high education and low wages will export IQ. That will be the Baltics, India, China. Of course, it's terrible if you're in Ohio or England or France or someone competing with an Estonian. So what you'll get is a massive mean reversion of income throughout the world where the Valley, Israel, China, and maybe a couple of other places get very high economic returns and everywhere else in the world starts to revert to the mean. Again, it starts to look more like the Roman Empire."

It is also the case that while the powers that be in Silicon Valley might not be the earliest movers in fields like precision agriculture, once success is achieved elsewhere, they don't just sit back as passive bystanders and watch it grow. Google chairman Eric Schmidt recruited an Israeli entrepreneur, Dror Berman, to move to Silicon Valley and head up Innovation Endeavors, a large venture firm that invests Schmidt's money. Israel is home to many of the 20th century's great innovations in farming. Berman brought the intellectual curiosity about agriculture with him to Silicon Valley and developed Farm2050, a partnership that aspires to combine data science and robotics to improve farming with a group of partners as diverse as Google, DuPont, and 3D Robotics. Dror recognized that Silicon Valley can be a little too navel-gazing, and told me that 90 percent of the region's entrepreneurs focus on 10 percent of the world's problems. With Farm2050, he is trying to bring Silicon Valley's A game to agriculture. Silicon Valley's history as a home to apricot and plum orchards is long past, and if it does establish itself as the source of winning investment or innovation for precision agriculture, it will contradict the idea that domain expertise will drive the industries of the future. Instead, it would suggest, as futurist Jaron Lanier has argued in his book *Who Owns the Future?*, that those who hold the most data, the fastest servers, and the most processing power will drive all growth from here on out. It's basically the idea that Google could do my job and your job—and everyone else's job—better if they wanted to simply by applying their top-of-the-line analytics abilities.

There is an increasingly large audience, however, that holds a different view from Charlie Songhurst. They believe that big data, instead of absorbing and supplanting other industries, will serve as a broad tool that every existing industry can use to spur growth. The idea is that data will become widely usable and scalable enough that it won't have domain expertise in the same way that other high-barrier-to-entry industries of the future like genomics or robotics do.

This view was explained to me by Mark Gorenberg, a veteran of West Coast venture investing who saw the investment case for analytics early and built a venture capital firm around it, Zetta Venture Partners. Mark has been in venture capital for a quarter-century and splits most of his time among investing, work with MIT, and serving as an advisor to the president of the United States as a member of the President's Council of Advisors on Science and Technology. Gorenberg believes the big data economy will extend far beyond Silicon Valley. He says: "Analytics businesses will come from anywhere. You have the algorithmic expertise on one side that is coming out of a lot of universities and you have the domain expertise for particular industries, which manifest themselves everywhere."

Gorenberg argues that as the big data market grows over the next couple of decades, it can be a source of revitalization for old industrial centers where local domain expertise exists. In the rust belt, for example, he sees strong opportunities for the development of analytics firms rooted in the region's strength managing industrial processes. He sees Boston's strength as a biotech center enabling it to create health data companies, and he foresees the creation of energy analytics companies in Texas. He predicts that we will see strong privacy and forensics companies forming around Washington, D.C., that build on the capabilities in the law enforcement and intelligence communities, providing high-paying jobs for people who worked at the NSA, CIA, and FBI.

If Gorenberg is correct that "domain expertise is everywhere," then there is no reason not to be optimistic about the prospects for big data firms developing outside the United States. It just takes a combination of algorithmic expertise and domain expertise. For example, having lost out on the wealth creation built around the Internet, Germany is now determined to leverage its domain expertise in logistics and household appliances to own the analytics markets in those areas of traditional strength, an initiative it calls Industrie 4.0.

If the big data market develops as Gorenberg predicts—with the best companies being headquartered all over the world—then wealth creation will look entirely different for big data than it did for the Internet, where the benefits were concentrated in a 30-by-15-mile area.

I was persuaded by Gorenberg's view when I was in New Zealand and saw an example of how the combination of big data and domain expertise will determine the geography of the industries of the future. New Zealand is home to twice as many dairy cattle as human beings. The Kiwis know cows. While there, I learned about the impact of Pasture Meter, a precision-agriculture technology developed in Palmerston North, a community of 82,000 people in the Manawatu-Wanganui region of New Zealand's North Island, more than 10,000 kilometers away from Silicon Valley. Pasture Meter uses advanced sensor technology to take 200 measurements per second over vast swaths of farmland to identify how much grass is in a paddock so that dairy cows can be distributed most effectively for feeding. It alerts farmers to the amount of feed they have and identifies low-production areas that need intervention from the farmer, say, more fertilizer. Traditional technology for evaluating fields, like ultrasound or rising plate meters, typically captures only 250 readings of a pasture, whereas Pasture Meter can take up to 18,500 readings. Anyone with a phone can use the technology, and it works regardless of factors like weather.

It may seem that monitoring pastures—literally watching the grass grow—is an unnecessary use of real-time analytics, but the Kiwi farmers know better. With the massive increase in upward economic mobility in China, there was increasing demand for beef and dairy products, but New Zealand's cattle farmers needed higher levels of efficiency—greater scale and lower prices—if they were going to sell into such a large market. China's population is 288 times the size of New Zealand's. With strong domain expertise in dairy farming, local farmers and farm-equipment manufacturers knew that if they could feed their cattle more efficiently, that would increase output enough for them to export to China.

What happened? Sales of beef from New Zealand to China soared 478 percent in one year. China surpassed New Zealand's neighbor Australia as New Zealand's largest export market, more than twice the size of what is exported from New Zealand to the United States.

The fact that it surprised me that it could be so important to know the location and concentration of grass in a field is exactly the point: New Zealand's farmers had the domain expertise, so they knew what needed to be built and they built it. It's too much to say that Pasture Meter deserves credit for the full 478 percent increase in beef exports, but local farmers bring it up as one of the important factors.

What happened in New Zealand can and will happen in other industries where there may not be a deep history of big data and analytics, but where there is domain expertise in another industry that knows where and how analytics would add value. The big data applications themselves are easily scalable, can be done broadly around the world, and can be implemented whether or not there is much previous data experience—as was the case for the people in Palmerston, New Zealand, who made equipment for the region's dairy farmers.

Silicon Valley builds things that Silicon Valley wants, from nicer taxi services to more photo-sharing apps. But investors and entrepreneurs in Silicon Valley don't see the world through, say, the eyes of farmers. Thus, they are less likely than a company in the Manawatu-Wanganui region of New Zealand's North Island to recognize the need for and develop a technology that enables greater beef and dairy production for export to China.

While Marc Andreessen is as closely identified with Silicon Valley as anybody, he agrees that those fields that are early in their development can and should take root wherever there is deep knowledge about a specific area. He has proposed that Detroit leverage its expertise in automotive mechanics to become "Drone Valley," and suggests that rather than trying to create more Silicon Valleys, we should hope and plan for the creation of "50 different variations of Silicon Valley, all unique from each other and all focusing on different domains."

Larry Summers reinforced Andreessen's view, telling me, "My general line is, in essence, there is much more division of labor than there used to be. Strategies for countries, companies, and people are much more about building on your strength than offsetting your weaknesses than it used to be." This effectively means to stop trying to chase after Silicon Valley and focus on the skills and processes that will unleash the next wave of innovation in fields in which there is already local expertise.

My view is that the geographic spread of domain expertise in the industries of the future will ensure that the next stage of globalization produces centers of innovation and commercialization that are more geographically diverse than the last stage, when Silicon Valley enjoyed 20 years of domination. There won't be a Roman Empire. The thought that software and big data–savvy companies and entrepreneurs in Silicon Valley will reign supreme is not a crazy one, but I think that as big data becomes more widely adopted, it will be more of a commodity that any industry can use. There is a real opportunity for stakeholders with domain expertise to innovate for themselves, and in breathtaking ways. But if they wait too long, some 28-year-olds in California are going to do it instead. In the cases where an industry is too slow to adapt, then eventually more efficient, less expert start-ups (with big data expertise) like Uber will step in and take down companies with decades of domain expertise. In the famous words of H. G. Wells, "Adapt or perish."

CITIES AS INNOVATION HUBS

The geographic foci for innovation are almost always cities. Why are cities growing so rapidly even as network technologies allow us to be more distributed, to do more at a distance? Three percent of the world's population lived in cities in 1800. Today 54 percent of the world's population lives in cities, and just 100 cities account for 30 percent of the world economy.

In some respects, cities have always been drivers of a society's growth, even when 97 percent of the population lived in rural areas. Empires have always been powered by their cities. Baghdad led the Abbasids to greatness. Rome did the same for the Romans, as did Constantinople for the Byzantines and then the Ottomans. The British colonized and established a chain of cities that linked their empire together, including Cape Town (colonized in 1814), Singapore (1824), and Hong Kong (1842). Today these key cities serve as their respective countries' and regions' links to the world, much as they did for the British Empire.

Cities are incubators of growth because they produce positive externalities, or spillover effects. They allow ideas, labor, and capital to flow rapidly and efficiently. Talent can be more effectively coordinated, and markets can become more specialized.

The most important cities from an economic standpoint are so-called alpha cities, for example, Shanghai, London, New York, and Tokyo. Exporting advanced services across the globe, they represent mini-economies in themselves. I am much more likely to run into an American who is a leading global innovator in an alpha city like Shanghai or Dubai than I am to see her or him in St. Louis, Missouri, or Manchester, England. They visit the same 20 cities around the world in a circuit of sorts, including San Francisco, Hong Kong, Singapore, Seoul, and Tel Aviv, to name some more of them. This is what globally engaged cities aspire to be.

Second- and third-tier beta and gamma cities link smaller regions. Berlin, Frankfurt, and Munich are not economic powerhouses by themselves, but they do contribute to a regional network that drives German national success. Each fills a particular service niche, and together they lead Germany to prosperity.

An important aspect of what makes major cities thrive is infrastructure, along with the analytics programs that allow people to use that infrastructure more efficiently. What makes a happy municipal citizen? Things like knowing exactly when the train will come, being able to go online instead of standing in line to access a government service, and being able to provide real-time feedback that informs how and when city services are delivered. The capability for doing this skews toward very large cities and increasingly involves big data, according to Stephen Goldsmith, a professor and director of the Innovations in American Government Program at Harvard's Kennedy School of Government. He directs Data-Smart City Solutions, a project "focused on government efforts to use and blend new technologies, big data analytics and community input."

As cities get tech savvy, this effectively means that global centers with lots of money and the capabilities Goldsmith describes (like New

York City, Dubai, London, San Francisco, Tokyo, and Seoul) are those likely to build big data applications that are highly appealing to citizens, and this attracts the "next economy" class.

Cities that are aspiring to become global hubs, like Jakarta, São Paulo, and Mumbai, need to simultaneously invest in physical infrastructure and the big data applications that often attach to this infrastructure. This helps create the conditions that attract investors and entrepreneurs.

Dubai in the United Arab Emirates and Doha in Qatar have used oil wealth to make themselves global hubs. In May 2014, Doha opened a $15 billion airport it hopes will compete for the title of world's busiest airport. The airport is one of the highest-tech places imaginable. With an expected capacity of 50 million passengers per year and the ever-growing traffic among Europe, India, and China, Doha has positioned itself as a hub of global people and capital flows.

But is this all it takes? Infrastructure is important, but what drives much of the growth in the world's leading cities? It turns out that they share a common culture of openness, even those in countries where the official government line is much more closed. The cities that are moving forward in the global economy are the ones most open to the outside world. Cities that have historically been open to the world are linked by a culture that welcomes people from all corners of the globe and encourages the free flow of ideas and goods to make them attractive places to live and work. The combination of first-class infrastructure and a high-density population gives us open access to other people: it's easy to communicate, gather, and exchange. Economic openness does the same on a business level, getting rid of friction that costs time and money to get business done. Political openness is an even higher-level mode of efficiency, ensuring that everyone in a society can gather, meet, work, and speak without the friction of undue censure or discrimination. From these increasing degrees of openness, we get a hint of what it will take to get a foothold in the industries of the future, even for countries and regions that have no alpha cities.

A BLEAK LANDSCAPE

Though the world's major cities are fueling the global economy, one does not need to be in an alpha city to succeed. In fact, Internet technology allows people to be almost anywhere and operate a successful business. But the technology itself isn't a cure-all; many limits can still hold even the most eager entrepreneur back.

Far away from any of the alpha cities is Waziristan, in the northwest of Pakistan, bordering Afghanistan. The landscape is bleak— rocky and gray—and the territory is virtually lawless. The Taliban roam about, carrying guns and enforcing a particularly harsh form of sharia. Drones buzz overhead. Pakistani army checkpoints dot the unpaved roads. Many villages are ghost towns, abandoned after years of fighting.

To be a woman in Waziristan is to be confined. You live in a small "women-only" section of your home. You can go outside only if a male relative accompanies you. Even if you make it outside, you can't do much when you are there. In the most recent election, mosque loudspeakers warned that women could not vote.

Somehow, emerging from this backdrop is Maria Umar.

The first time I met Maria was in Islamabad, Pakistan. She wore a traditional Pakistani *shalwar kameez.* It was bright magenta—a splash of color in a sea of dark Western suits. Her hair was pulled back in a loose ponytail. Maria talks fast, laughs easily, and uses social media obsessively. She is an entrepreneur who has built an impressive company by putting social media and Pakistan's most underused resource, women in the farthest reaches of the country, to work.

Maria's family is from Waziristan—or, as she puts it, "Yes, my parents, my in-laws, myself, we all belong to this place called Waziristan." She now lives in Lahore, where her husband is stationed as a civil servant, but no matter her physical location, the virtual business is bustling.

Maria Umar shares her story on a Skype call from her son's

bedroom in Lahore, the only place in her home where she can keep her kids out and have a quiet conversation. A Barcelona soccer flag is draped on the wall above her son's twin bed, which she sits on as she talks about her improbable success.

The story starts six and a half years ago when Maria was pregnant with her second child and teaching in a local private school. The school could not give her maternity leave, so she was let go.

"At first I was excited. I thought, okay, I can have a nice break after three years of working and I will go visit my family, I will watch TV," Maria says. "But within a week I had called everyone I needed to and watched all the TV I wanted to watch. And then what next?"

She had heard people talk about work opportunities online and decided to look into it herself. Maria found a freelancers' platform and started applying for jobs. "It was a lot of hard work just getting that first task," she says, "because all the tasks were going up mostly in Eastern Time, and here I was in Pakistan."

Maria's computer was in her and her husband's bedroom, and in those early days she would keep him and the rest of the family up as she raced to finish projects in the middle of the night, aligning her schedule with those of her US-based clients.

Those first projects were for blogs. She wrote a post about a hair straightener—and had to rewrite it nine times to make it "unique." From her home at the time in Rawalpindi, Maria blogged for a baby boomer website about everything from religion to dream interpretations to aging gadgets. "It just started off that way," Maria says. "And then I started getting more work than I could manage, and I started giving it out to my cousin, my niece, and their friends."

The same clients she had been writing for started asking Maria to take on other services—basic Internet research or maintaining their Facebook page—which Maria sent along to her network.

Initially Maria's network consisted only of friends and relatives, all of them women. In fact she hadn't realized all her projects were going to women for several months. But once she did, "I saw the

impact both on them and in my own life," she says. "It made me realize just how potent a catalyst for change this [work] could be in a
country like Pakistan where literacy rates are on the rise but there are
still obstacles to women working outside their homes."

Two years after she took her first freelance assignment, Maria's
business was booming. This young woman from Waziristan was putting more and more Pakistani women to work for American clients.

"A mentor of mine said that unless you put a name on it and
print business cards and start introducing yourself as 'Maria Umar,
Founder-President of Women's Digital League,' it's never going to be
a serious business—neither for you nor for anyone else who listens
to you," she says. So she printed business cards and ramped up her
business. Despite normally being between nine and twelve time zones
and thousands of miles away from her US clients, she was making a
name for herself. Maria used the American networking site LinkedIn
to connect with her clients and asked them to endorse her work on the
site. "It was through these endorsements that I would get more work."

As more jobs came in, she updated her company's LinkedIn profile
with the newest jobs and skill sets. LinkedIn became "a place where
I can post a brag post about myself. If we completed a big task, we'll
post that up there," she says.

Clients hire Maria—and ultimately Maria's large network of
freelancers—to do computer-related tasks. An e-commerce group
needs its logo put on a bunch of photographs. A food website wants a
list of all sushi restaurants, including addresses, in a certain zip code.
A company needs a Facebook app developed.

When a job comes in, Maria does not turn to LinkedIn. She explains that in Pakistan, the site is "not really for hiring people, that's
more of a Facebook, Twitter thing," she says. She posts the job announcement for her network with hashtags like #homebasedwork and
#writerneeded. She gets a new résumé every day and has 500 women
in her freelance network now. "We're connected to people who are really good at what they do," she says.

Her business, she says, has always been about women's empowerment. Most Pakistani women marry in their early twenties and have kids soon after, Maria explains, so "full-time jobs are just not possible for most. The Women's Digital League is a platform for these women to find a whole new way of working."

Maria hopes this is just the beginning. "The idea is to spread out to the Middle East and to the Arab region," she says with a wide smile. "The way the women are treated or the challenges they face in the workforce are kind of similar around these areas, and I think that something like this online platform—where women with different skill sets, from the very basic to more advanced, can come together—can really make a difference. It's an emotional difference rather than just a financial difference."

The 36-year-old entrepreneurial success from Waziristan concludes, "I have seen how huge a difference it has made in my life and I would like it to spread out."

THE 21ST CENTURY IS A TERRIBLE TIME TO BE A CONTROL FREAK

Maria's story is a case study on the importance of openness. She was only able to succeed because she had open access to the Web, which enabled her and the women on her team to connect and do business with people all over the world while working from a restrictive environment in Pakistan. However, she had to work triply hard to succeed because there was no political and little economic openness, which shows that the countries that are going to be able to succeed economically won't be able to do so by being control freaks.

I saw this in another part of the world when I was on a diplomatic mission to Russia to discuss innovation and Internet freedom. I went to meet with Vladimir Putin's top political and communications aide, Dmitri Peskov, at the Russian White House (the Russians call their government headquarters and prime ministerial residence the "White

House" too). My team was led to a very formal room with Peskov and his staff at a long rectangular table. On one side were eight Russians, on the other side eight Americans. Peskov and I sat at the middle of the table and did most of the talking during the 90-minute meeting. A couple of times during the meeting, Peskov stood up and walked through one of the doors—the one that led to Vladimir Putin's office—and returned a few minutes later with fresh questions.

The most remarkable moment came when I asked Peskov about the power of the Internet to reduce corruption, a rampant problem in Russia. He dismissed the idea that the Internet was a reasonable place to name and shame corruption: "It's the Internet people who are corrupt!" He then held up the pen sitting on the table in front of him for everyone to see. "I can prove to you that using the Internet makes you forget how to use this," he announced, waving the pen in front of me. "I can prove with research that it makes you mentally handicapped!" Peskov believes that Internet use leads to a loss of vocabulary and measurable intelligence, a view in keeping with the Russian government's official position. Putin refers to the Internet as a "special CIA project." He reportedly has never sent an email.

As we walked out after the meeting with Peskov, one of his assistants, a woman with bright red hair who was probably six feet tall, peppered my deputy, Ben Scott, and me with questions about our office—what we did, its size and its composition—in such detail that we ended up laughing. At one point, she would not open a door until we offered an answer to a question about how many people worked in my office. That's how the Russian government now conducts business and manages its politics, society, and the economy.

Since first taking office in 1999, Putin has reimposed control over the media, eliminated the freedoms of association and assembly, divvied up control of the economy to his chosen hangers-on, and concentrated power in his office. His policy is part of a longer Russian history of autocratic leadership that extends back centuries. Russia has always suffered from a schizophrenic relationship with the outside world. A zeal for control from the top has always been at odds with the require-ments for more openness. While leaders like Peter the Great, Cath-erine the Great, and Mikhail Gorbachev wanted to open Russia up to new ways of thinking and doing business, most of the czars and Soviet premiers, and now Putin, have wanted to impose control not only over Russian politics but also over Russian society, the Russian economy, even Russian thought. Putin, a former KGB agent, is symp-tomatic of this trend in Russian history.

Putin's paranoia about the outside world is itself at odds with the requirements of a modern economy. To be successful in the next wave of globalization and innovation, a society must be open in order to ex-change new ideas, conduct research free from political interference, and pursue creative projects, even if they fail. Innovation requires this stripe of openness. It cannot see outside markets as enemy territory. During the brief period when Putin stood in the background and Dmitri Med-vedev served as president, Medvedev took a shot at creating a Russian version of Silicon Valley in Skolkovo, outside Moscow. The Skolkovo Foundation followed the "how to build a Silicon Valley" template Marc Andreessen laid out line by line. A total of $2.5 billion was invested and a laundry list of corporate partners lined up to participate, includ-ing Microsoft, IBM, Cisco, and Samsung. But Skolkovo foundered once Putin returned to the presidency and any hope of a culture lending itself to innovation was lost. In a notable contrast with Silicon Valley's streak of libertarianism, Putin's crackdown included discrimination against homosexuals, and women now face legal restrictions in 456 specific jobs. The engagement by non-Russians in Skolkovo all but disappeared.

Putin does not understand, or does not care about, the basic real-ity of how growth in the global economy is now produced. Locked in a 19th-century mind-set in which land, power, and people are physically controlled, Putin is missing the reality of power in the 21st century. The nature of economic success is different in an information-based economy than in an industrial or agricultural economy, where iron and land are king.

BREADLINES AND BROADBAND

Rarely do countries and societies have the opportunity to make a simple, binary choice about whether they are going to be open or closed. But that is exactly what happened after the dissolution of the Soviet Union and the reestablished independence of Estonia and Belarus. The two countries are separated by just a few hundred kilometers west of Russia, but their trajectories could not be more different.

Estonia is "The Little Country That Could," the title of a book by the first prime minister of Estonia, Mart Laar, which explained the country's rise from ruin at the end of Soviet occupation in 1991 to become one of the most innovative societies in the world today.

Following Estonia's independence with the collapse of the Soviet Union, its economy was left reeling. Everyday life for most of the country's citizens was dire. Its currency was stripped of any value. Shops were empty and food was rationed. The gas shortage was so bad that the tattered government planned to evacuate the capital of Tallinn to the countryside. Industrial production dropped in 1992 by more than 30 percent, a higher decline than America suffered during the Great Depression. Inflation skyrocketed to over 1,000 percent, and fuel costs soared by 10,000 percent. The only system left working was the informal market, which, along with weak legal protections and unprotected borders, facilitated an increase in organized crime for Estonia and its neighbors. This was happening right around the time that Silicon Valley was about to take off with the advent of the commercial Internet.

Led by the then 32-year-old historian Mart Laar, elected prime minister in 1992, the new Estonian government wasted no time charting a new course. "To get my country out of this mess and collapse," he explained, "demanded radical reforms—as most medicines they were unpleasant at the beginning."

Laar's first step was to stabilize the economy. Because it couldn't yet print money and had no effective mechanism to raise cash, the government cut expenses, particularly to traditionally insulated sectors. In ending subsidies to state-owned companies, Prime Minister Laar announced that companies must "start working or die out."

The focus on fostering an innovative business culture rooted in openness was reflected in the government's second stage of reforms. After the economy had been stabilized—inflation had dropped from 1,000 percent in 1992 to 29 percent in 1995—Estonia opened its doors to the world economy. It reduced trade tariffs and ended all export restrictions, turning the small country into a trading hub. The government also reached out to foreign investors. The citizenship law was amended to provide equal civil protection to resident aliens. Estonia passed laws to ensure that foreigners could purchase land. All special privileges for existing investors, many of them holdovers from the Soviet era, were abolished in order to ensure a level playing field for new investors.

When the Soviet Union collapsed, fewer than half of Estonians had a telephone line. The Finnish government, in a gesture of philanthropy, offered Estonia its analog phone system for free as the Finns upgraded to a digital network. The Estonians declined, choosing to bypass analog telephony and move straight to a digital network of their own design. As it developed its own government, it skipped the typewriter-and-paper stage and began putting its services online from the outset. Every school in Estonia was online by 1998, just four years after the birth of the commercial Internet and six years after widespread fuel shortages and breadlines. In 2000 Internet access was legally enshrined as a human right by parliament.

Estonia quickly became a center for global investment. The country received more foreign investment per capita in the second half of the 1990s than any other central or eastern European economy. This investment allowed Estonia to upgrade its technological and industrial base, and laid the ground for an innovation economy.

Since independence, Estonia has been led by technocratic governments that have further opened the economy. In 2008, under President

Toomas Ilves, Estonians voted to join the European Union (EU) and have since adopted the euro. As part of the EU, the Organization for Economic Cooperation and Development, and the World Trade Organization, Estonia ranks as one of the most globally integrated eastern European economies.

President Ilves credits the impetus for the reforms that Estonia has enacted to its "willingness to actually do things differently and a huge amount of political courage. We enacted just a clean privatization process, basically following the German Treuhand model: quick reform of the tax system, introducing of a flat rate income tax. We computerized and were the first [former Soviet] country to establish its own currency. And we did that against the advice of the IMF, so it was actually a fairly visionary approach by the very young people who were elected in the general election of 1992. If you go back into the late, very late Soviet period, the Estonians kept coming up with all kinds of proposals for reform that were largely shot down, but nonetheless existed. There was, at least in the late 1980s and early 1990s, a spirit here that we can do things, and we'll be brave and do things differently."

The strategy was one of near-radical openness combined with an orderly and disciplined framework for how and when to open up the economy. The result is that Estonia has achieved a standard of living far beyond that of 20 years ago. Its GDP of over $25,000 per capita is 15 times what it was at the fall of the Soviet Union and ranks number one among the fifteen former Soviet republics.

The real success of Estonia is reflected not only in these statistics, but also in its place as one of the world's leading centers of innovation. Estonia has not produced a centi-billion-dollar company like Google, but it has achieved some notable successes, including Skype. More significant, it has innovated in a way that every place in the world, including Silicon Valley, should envy. In doing so, it has improved its civic and political life in a way that positions it as well as any place in the world for the industries of the future.

CLOSED FOR BUSINESS

Estonia and Belarus were in nearly the same position following independence and made opposite decisions about their future. While Estonia opened up, Belarus closed off.

Under Alexander Lukashenko, who has ruled since 1994, Belarus has maintained a tightly controlled political and economic system. Lukashenko is the ultimate control freak. He runs Belarus as his personal fiefdom. Dissidents are silenced. The press is tightly controlled. Joining an opposition protest may result in being labeled a terrorist. Opponents should expect that Lukashenko will, in his own words, "wring their necks, as one might a duck."

In economic terms, Lukashenko is a neo-Luddite—someone who simply does not get the modern world. He owns the economy as much as he does the political system, even if the Belarusian economy does not amount to much. A former farm manager, Lukashenko is the single key player in its economy. Most businesses are state owned—effectively Lukashenko owned—and output and employment are subject to strict administrative controls. Around 40 percent of industrial enterprises and over 60 percent of agricultural firms incur losses. Belarus's currency is the Belarusian ruble, which makes the Russian ruble look strong by comparison.

Belarus is still a land producing practically no data. It is a remnant of the 1970s, with typewriters still in use in a large percentage of businesses and government offices. Instead of using "serf-like" robots to replace manual labor, Belarus is stuck in an era where people are still effectively serfs. Belarusian laborers still toil on collective farms or in outdated industrial manufacturing. They do the dull, dirty, dangerous jobs that robots are doing in more advanced economies.

The high-water mark for Belarus and the Internet is a social media–savvy graduate student in Massachusetts named Evgeny Morozov, who writes neo-Luddite screeds against American technology companies, advancing the official views of Russia and Belarus.

President Ilves of Estonia explains, "I don't think there was that big a difference in '91 and '92 between the two countries, but then autocracy takes its toll, and they didn't undertake reforms."

When I first arrived in Estonia and drove into the capital city of Tallinn, I noticed that the headlights on our car lit up reflectors on all the pedestrians we passed. Bracelets and necklaces were lighting up like the stripes on the vests of road workers working at night. One of the first people I met was Karoli Hindriks, the CEO of Jobbatical, a company that blends the concept of a job and a sabbatical, matching employers and talent for short-term jobs that might involve sending a software developer from Sweden to Thailand for a three-month "jobbatical." I asked Karoli why she and everybody else on the street was wearing reflective clothing, and she told me that when it becomes dark, it is the law in Estonia that all pedestrians wear some form of reflective clothing for safety reasons. She smiled and told me that she became an inventor at age 16, creating pedestrian reflectors that could be used in clothing and jewelry, and she now holds several patents and international trademarks for her designs.

This was representative of what I have seen during all of my time in Estonia: extreme order combined with invention and design.

President Ilves is not quite like any other head of state I have encountered. He has a distinct look (a hip buzz cut with three-piece suits and bow tie) and interpersonal style—one part tough guy, one part technology geek. Ilves was raised in the diaspora (New Jersey, actually) and returned to Estonia after the fall of communism and Estonia's reestablished independence from the Soviet Union. In the hundreds of thousands of miles that I traveled for the State Department and the many meetings I had with foreign dignitaries, Ilves struck me as the most knowledgeable of the world's 196 heads of state on technology issues.

Today Estonia is one of the most connected countries in the world. It has the world's fastest Internet speeds and universal medical health records, something that the United States has been struggling with for years, with no end in sight. In 2007, Estonia became the first country to allow online voting in a general election. Ninety-five percent of Estonians file their tax returns online—doing so takes about five minutes.

In December 2014, Estonia made yet another bold move, offering what it calls "e-residency" to any person in the world. As the country put more of its government services online, from incorporating a company (which happens at a world-leading speed, estimated at five minutes) to authenticating electronic signatures, it has seized the opportunity to position itself as a hub for digital government services. To become an e-resident of Estonia, you make one trip to the country (though it hopes to be able to operate out of its embassies in the future) to submit your biometrics and other personal data for verification. You pay the registration fee and receive a secure chip-enabled identity card. You can now use your Estonian e-residency for a variety of things, such as doing business throughout the EU and leveraging its online-only programs for contracting and tax filing. It's a way to bypass other countries' more expensive and less efficient systems. No more paperwork, lower taxes, and, if you own a business, all the freedom that comes with being an incorporated business in the EU. In a similar way to how other countries have created tax havens to benefit from large deposits in their banks, Estonia has established itself as an efficiency haven. Instead of facilitating criminal behavior as tax havens do, Estonia's system is trying to make business more secure. The ideology behind it is rooted in good government. Among the benefits to Estonia is the additional tax revenue and more than $500 million in fees alone that it expects from 10 million e-residents over the next several years. Every leader I've spoken to about Estonia's e-residency has the same one-word, three-letter response: *wow*.

Estonia is putting its newfound prosperity to good use. The country now spends a larger percentage of its GDP on primary school education than the United States, the United Kingdom, and nearly every other European country. School enrollment and literacy are at 100 percent. All schoolchildren are taught how to code beginning in the first grade. President Ilves explained to me that to be competitive in tomorrow's economy requires his country "to change our educational system in a way that people who go through it come up with skills that are in fact useful in a robotic and computerized and automated era. . . . This is one of the reasons why we want to teach all kids from grade one how to program. I mean we already start teaching foreign languages early. Computer language is just another language with its own grammar; it just happens to be much more logical than French."

Ilves thinks that the advancement of robotics serves Estonia well by giving the small countries of the world the chance to compete on the global stage with actors like China and India. He told me, "It will increase our functional size tremendously because people don't have to do things the machines can do." Estonia has only 1.3 million citizens. Ninety-eight cities in China have a larger population than the entire country of Estonia. At the core of Ilves's thinking is the fact that robots will enable significantly increased output per capita. How does a little country like Estonia compete in the same global marketplace as China, which has a labor force a little more than 1,000 times the size of its own? It takes advantage of the fact that robots enable a relatively small workforce to produce higher levels of output than would be the case in an all-human workforce. Estonia and China will never be equal competitors by sole virtue of their difference in size, but Estonia can compete at a level far above what its size would suggest by virtue of being cutting-edge in the field of robotics as both producer and consumer.

Estonia has demonstrated how innovation in the industries of the future can do more than just generate wealth and employment; it can enhance our civic and political life. In this respect we should stop asking about the next Silicon Valley and start asking about the next Estonia.

UKRAINE: THE CLASH BETWEEN OPEN AND CLOSED

The choices that Belarus and Estonia made 25 years ago, when both had breadlines and broken economies, are representative of the choices being made about political and economic models around much of the world today. The choice of whether to adopt an open (Estonian) or closed (Belarusian) economic model is at the heart of the conflict in Ukraine.

Ukraine has long sat on the dividing line between geopolitical forces: the West and the East, Europe and Russia, Catholicism and Eastern Orthodoxy. Ukrainians in the west of the country traditionally tilt toward Europe, while those in the east of the country (the majority of whom are ethnic Russians) tilt toward Moscow. The very name *Ukraine* means "borderlands."

In Silicon Valley on February 19, 2014, the day after the protests broke out in Kiev, Ukrainian American WhatsApp founder Jan Koum signed a $19 billion deal to sell his company to Facebook. For Ukraine, that same $19 billion would have been the answer to its short-term bond, debt, and gas bills.

The fact that Ukraine's economic lifeline could be equal in cost to the purchase of a mobile messaging app—created by a Ukrainian emigrant—exemplifies how much potential Ukraine has and how badly that potential was being squandered under Ukraine's prior, Russian-model government. Koum, born in a village outside Kiev, emigrated from a politically unstable Ukraine as a teenager, and a bright, innovative mind that could have helped uplift Ukraine instead ended up elsewhere.

At the turn of the 20th century, my own Kiev-born great-grandfather made a similar choice. Disillusioned with authoritarianism, he became an anarchist, a decision that forced him out of the country. He eventually made his way to Chicago, and it was there, instead of in his native Ukraine, that he ended up starting a small business and settling down.

The last 100 years saw too many people like my great-grandfather and Jan Koum, desperate to get out of Ukraine—or other countries that felt more stifling than encouraging.

While working at the State Department, I saw the technological savvy of Ukraine's young people. Ukraine is the number one outsourcing destination in central and eastern Europe for information technology services. Its notable technology entrepreneurs include more than just Jan Koum. Enable Talk, a project described in chapter 1 that uses special gloves to translate sign language into speech, was created by four Ukrainian student developers. Enable Talk took home the first prize at the Microsoft Imagine Cup competition in 2012, and *Time* magazine named it one of the best inventions of the year.

PayPal cofounder and serial entrepreneur Max Levchin also comes from a family that fled Kiev to seek political asylum in the United States. Science and technology companies in Silicon Valley, London, and Berlin are teeming with Ukrainian engineers. Ukraine's black hat hackers-for-hire are some of the best in the world.

At the very moment Koum and Mark Zuckerberg were finalizing their deal, female entrepreneurs in Ukraine were preparing for an event called Startup Weekend Kyiv. Soon after the breakout of protests, the group's website read: "Due to political turmoil this event has been postponed." Mired in corruption, kleptocracy, and authoritarianism, Ukraine has not nurtured the Koums of its future. As Koum tellingly tweeted about his adoptive homeland in March 2013, "WhatsApp Messenger: Made in USA. Land of the free and the home of the brave."

When the pro-Russian government was swept out of power during the Maidan protests in 2013, Ukrainian citizens were trying to create the conditions that would keep Ukraine from losing out on the economic benefits that companies like PayPal and WhatsApp could have created in the last 20 years.

The president who succeeded Ukraine's pro-Putin government, Petro Poroshenko, recognizes that it can be home to multibillion-dollar breakouts if it creates a functional environment for its innovators. Po-roshenko himself is a rich businessman and owns Ukraine's largest confectionery manufacturer. The main slogan of his election campaign was "a new way of living."

I have gotten to know Poroshenko fairly well. He and another Ukrainian oligarch, Victor Pinchuk, were considered "good oligarchs" by the US government. Putting aside how they made their money (which I do not entirely understand and which I don't think anybody other than Pinchuk and Poroshenko themselves really do), the perception is that these two want Ukraine to move to more openness, less corruption, and a more Jan Koum–friendly business environment. I think it's possible, and I choose to think of myself as hopeful rather than naive in that respect.

Poroshenko and I have had many conversations about the industries of the future and the future of Ukraine. He was usually fingering prayer beads during our discussions. That seems appropriate.

THE CHOICES

Estonia and Belarus are two poles on the open-closed axis. Most of the world lies between them, and many countries, like Ukraine, are torn between the two. States as different as Turkey and Thailand regularly go through public convulsions as they try to reconcile diverging tendencies toward open and closed systems. The combination of global history, international competition, and a raft of local political variables is producing a series of hybrid systems across most of the world's countries. These include China's state-run capitalism; India's complex and sometimes dysfunctional democracy with a market economy that still has significant distortions; Western Europe's social market economies that are struggling to adapt to the pressures of austerity and aging; America's increasing political polarization over the role of the market in its economy and society; and mixed development strategies across Africa, Latin America, and Asia.

Everywhere political and economic systems face diverse but familiar problems: how to balance growth and stability in an era of rising inequality; how to prepare for the social and economic challenges that will come with the next wave of globalization and innovation; and how to become a base for that innovation and the corporate headquarters that will come from it, or at least to be in the supply chain for companies that are headquartered elsewhere.

The foundational question being asked and answered by heads of state is about how much control to exert over society. When I speak with leaders around the world and ask them what one thing has most changed for them over the past 15 years, they almost always cite the perceived loss of control. The thing they cite as the major reason? Technologies like the Internet and social media that connect people to information and each other.

Media and information environments, political agendas, social movements, governmental decision-making processes, and control over corporate brands have all been disrupted by citizens using what are now billions of devices and billions of Internet connections. Information no longer flows exclusively from mainstream media and government out to society. It flows in a vast network of citizens and consumers interacting with once-dominant information sources. This network of people is constantly reading, writing, and evaluating everything, shaping the ideas that guide society and politics. These connection technologies give power to citizens and networks of citizens that would have once been reserved to large hierarchies such as media companies and governments.

How states respond to this systemic loss of control and the diffusion of power will greatly affect the character and performance of their economies. The principal political binary of the last half of the 20th century was communism versus capitalism. In the 21st century, it is open versus closed. There are no 100 percent open states or 100 percent closed states. Openness can be applied selectively to the economy or to society and political systems more broadly. But to the extent that political systems are somewhat deterministic, countries that choose more open systems and can maintain them will be the places where the industries and businesses of the future are founded, funded, and come to market.

2.5 BILLION PEOPLE

Can the geographic centers for the industries of the future maintain openness even in restrictive societies? It's a struggle still sorting itself out. On the whole, socially and politically open countries have been better able to thrive economically. But some societies have managed to rise in recent decades by opening up economically and socially while restricting political openness. Whether this is a viable long-term strategy remains to be seen, but the rising powers that are now grappling with hybrid models are worth examining more closely, particularly Singapore, China, and India. By most measures, Singapore is one of the most innovative and economically successful places on the planet. It has a sky-high per capita GDP of more than $78,000, higher than the United States, Switzerland, and the oil-rich United Arab Emirates. There are all sorts of indexes that rank countries by how innovative they are, and Singapore is always in the top ten.

With just 5.4 million people, Singapore has the systems and responsibilities of a national government but the size of a big city. It places restrictions on free expression and assembly but is also the most religiously diverse country in the world.

The world's two most important rising economic powers, China and India, are both grappling with the growing need for openness in their own ways. Together comprising over a third of the world's population, they have achieved the most rapid breakthroughs in development in human history. Economic reforms have lifted half a billion people out of poverty in China and cut poverty by more than half in India. They have transformed from countries where famines killed tens of millions of people during the 20th century to two of the largest and most vibrant economies in the world. Their futures will see transformative change every bit as exceptional as the past three decades have.

For decades, China demonstrated that a somewhat open economy and a closed political system can achieve growth by being home to knowledge workers and manufacturing centers. But it is now seeking to prove that it can provide the conditions for innovation of its own. To this end, the core question for China's future is whether its model of relative economic openness but tight political control can foster real innovation. Thus far, it seems that its knowledge economy has been hampered. For example, China's successes in the Internet economy have all come from either building Chinese versions of technologies previously invented in the United States or Canada (and often stealing the intellectual property to do it) or from providing low-cost manufacturing to build the hardware for non-Chinese companies.

But while the control-freak impulse from Beijing has hindered the development of China's knowledge economy, it hasn't killed off the spirit of Chinese innovation. Jack Dorsey senses a level of vibrancy coming from China's entrepreneurs. He told me, "The people working at the companies have a feeling like they can create anything and really take on the world. On the other hand, when you talk to the government—and I'm just talking about intangibles here—it just feels a little bit more obviously controlled; it doesn't feel as excited. Whereas if you talk to the people, there is a real freedom in their speech. They see everything happening with Twitter or Square and they see that this could be everywhere, and I don't think they actually know how to start. But the feeling is definitely there. The energy is there."

The Chinese government may be slow to change, but they're also aware of the energy that Dorsey refers to. The government has realized that if the country is going to continue to grow, it can't merely be a center of low-cost manufacturing and copycat innovation. Having lost out on the wealth creation that came from being the source of innovation, investment, and commercialization for the Internet, they are determined to set the pace and be home to headquarters cities in genomics, robotics, cyber, and other industries of the future. China now aims to become a global center of innovation, encourage "internal market" development, and "rebalance" its economy to redress some of the social and environmental costs of its rapid modernization.

As growth has slowed in China, it has made moves to open itself up economically while keeping a lid on political openness. An early experiment in openness is the Shanghai Free Trade Zone (FTZ), an 11-square-mile business-friendly open economic zone. In the FTZ, the yuan is more easily convertible, there are fewer restrictions on foreign capital, and companies can import goods with lower trade barriers. One visible example is Microsoft selling its Xbox in the Shanghai FTZ, the first game console to be legally sold on mainland China in more than ten years.

Many Chinese and non-Chinese are withholding their enthusiasm, though, because the Chinese government has proved unwilling to loosen restrictions on foreign news sites and social media like Facebook and Twitter, as was originally rumored and reported with the announcement of the FTZ. The *People's Daily*, which the government uses to get its views to the public, shut down people's hopes when it reported, "The Shanghai FTZ is a special economic zone but not a special political zone. No one in their rational mind could imagine that the second-largest economy in the world, after over 60 years of striving, would set up a 'political concession' when it is thriving day by day."

The Chinese government's strategy is to jump-start development in seven key industries: energy saving and environmental protection, new-generation information technology, biotechnology, high-end equipment, new energy, new materials, and new-energy vehicles. First announced by the State Council as part of the twelfth Five-Year Plan in October 2010, these initiatives are intended to upgrade China's innovation capacity. Currently these industries comprise 4 percent of

GDP, but the Chinese leadership hopes to boost this to 15 percent by 2020.

Though a frequent critic of the Chinese government, Google chairman Eric Schmidt thinks China will maintain its current economic momentum: "On a pure economic basis, China has the strongest decade ahead of it. It is probable that Chinese growth slows in a decade and is succeeded by other Asian countries. We don't know about India; it is possible that their inefficient democracy will hurt them." John Donahoe, eBay's former CEO, agrees: "Fifteen years from now, China will, I think, be a very respectable global competitor."

China's neighbor India has transformed from the India of my youth, the country of Mother Teresa, famine, and a nasty caste system. Poverty is still grinding in parts of India. When I was at the State Department, I spent time in large slums walking through ankle-deep feces that reminded me of the worst of East Congo. But the prevalence of poverty is decreasing, and an emotionally detached look at the mathematics of India's economy shows an impressive 25-year run in its past and another promising 25 years in front of the nation.

India is a highly diverse country in linguistic, ethnic, and religious terms. Its citizens speak 780 languages. It has maintained a relatively pluralistic, secular democracy even as its unwieldy regulatory processes and complex government system have resulted in severe inefficiencies and market distortions. India has 29 states with constitutionally protected powers that in much of the rest of the world are reserved for national governments. Union lists and state lists define what the central government can and cannot rule on, which makes the Indian market difficult to navigate. The World Bank ranks it as the 142nd hardest place to do business out of 189 countries.

The contrasting effects of China's central planning versus India's more democratic and inefficient agenda setting are significant. China has developed as a manufacturing hub in no small part because the central government could plan it. It leads the world in infrastructure investment, a buildup that has directly contributed to its booming manufacturing sector, and it has enacted a forced urbanization policy to keep manufacturing wages low. Yet these policies have come with considerable human and environmental costs. As economist Nouriel Roubini describes, "In China, if they need to raze entire neighborhoods to the ground and push out millions of people every year from farms to support urbanization, they can do it against anyone's opposition because it's not a democracy. And because of what China has done, the environmental damage has been massive and widespread—air, water, land, food security, you name it."

India's lack of infrastructure has hindered its export and manufacturing sectors. Roubini observes that "in India you see the bypass of infrastructure. In Mumbai, flyovers go over shantytowns. A homeless person may have the right to not be moved from the little place he sleeps on the street. It can take years, then, to move stuff forward. This is why infrastructure is superdeveloped in China and is underdeveloped in India."

What India lacks in central planning for manufacturing it has made up for in producing knowledge workers. India trains around 1.5 million engineers every year, which is more than the United States and China combined.

India's first prime minister, Jawaharlal Nehru, focused significant resources on IT and higher education. His government oversaw the establishment of the All India Institute of Medical Sciences, the Indian Institutes of Technology, and the Indian Institutes of Management, which are among the best professional training centers in any emerging market—indeed, in any market at all. This talent base fuels foreign direct investment. Over the past two decades, many multinationals have shifted research and development departments to India. Call centers, medical billing centers, and other business administration services also developed rapidly. However, India has neglected primary education, leading to widespread inequality of opportunity.

The single best thing that could happen for India's positioning as a center for the industries of the future would be for Prime Minister Narendra Modi's government to make the kind of commitment to primary education that the Nehru government made to higher education.

Economist Rob Shapiro emphasizes the importance of education in building an innovation ecosystem. Citing China's and India's lack of education investments, he believes that in terms of innovation, neither is a contender for global power. This contrasts with tiny Singapore. China's population is 251 times that of Singapore, but they compete fiercely against each other in innovation. Singapore is able to compete with China because it delivers some of the best primary education in the world. My wife is an award-winning middle school math teacher. Her secret? Her program has been given permission to use Singapore's curriculum.

While Shapiro downplays China's potential to innovate, he does note that "you have to be open to the possibility that under certain conditions, the closed paradigm could be very successful." It's rare, Shapiro suggests, but it can function as a form of transition: "The single most successful country of the last half century is South Korea, which has achieved modernization and growth that China envies. And they did much of it under terrible dictatorships. Now, South Korea is a small country, and they were lucky to have some really smart dictators. They also made a peaceful transition to become a democracy, which is very unusual."

India's development has taken place from the opposite direction of South Korea's and China's: it has a much higher degree of political openness, and the development of the economy has taken place away from the command-and-control-like structure of the central government. Now it is trying to move toward more centralized control. India's new government was elected in no small part to take full advantage of all the engineering talent it is producing and wipe out its inefficiencies, perhaps convincing the likes of Eric Schmidt and John

Donahoe that India's growth will be the economic story of the next ten years.

Prime Minister Modi is trying to accomplish this in significant part by turning his country's 1.2 billion people into code; that is, he is racing to establish the world's largest biometric identity program, which would be used to deliver government services, subsidies, and information, Estonian style. The biometric cards are called Aadhaar, the Hindi word for "foundation." After the Modi government began distributing subsidies through Aadhaar, online bank accounts were created for 120 million households, pushing people into savings, loan, and payment systems designed for the future. As of this writing, 770 million Indian citizens, 64 percent of the population, had Aadhaars. The bet is that the kinds of systems that allowed Estonia to become the world's most innovative government can be applied to similar effect even in one of the world's most diverse and populous countries. India would clearly like to end up more like Estonia than Belarus.

196 COUNTRIES, 196 CHOICES

The future of the global economy depends hugely on what happens in China and India, but countries around the world are facing the same predicaments. Some are adapting in brilliant, innovative ways, while others are languishing or failing to realize the shifting winds of the global economy.

Latin America, for instance, is a patchwork. I saw countries that were slingshotting into the future—Chile and Colombia, for instance—where young technologists are building world-class companies. Others countries, like Ecuador and Venezuela, are stuck entirely in a dysfunctional past—in no small part because they have been led by controlling governments.

Brazil, which due to its size has the most potential of any country on the continent, is trying to develop its own model, but like India, it is struggling to get its own model right. In the 2000s, Brazil did an impressive job creating a path for more than 35 million people to move from poverty to the middle class, but the country is still being held back from greater growth by its high degree of neomercantilism— its lingering control-freak economic model that imposes stiff tariffs on imports and government control over economic activity involving non-Brazilians.

Brazil's neighbor and rival, Argentina, has been held back by control-freak economics as much as any country in the world. From 1870 to 1914, Argentina had the highest annual growth rate—at 6 percent—in the world, based on an extremely open economic model. Argentina primarily exported agricultural products—beef and wheat— from its rich heartland, the Pampas, in exchange for foreign invest- ment. Workers poured in as part of Argentina's nearly open (at least for Europeans) immigration policy. During the same period, Argentina received the world's second-largest number of immigrants, after only the United States, many of them Italians who chose Argentina over America. By 1914, Argentina ranked among the ten richest countries in the world, ahead of Germany and France. Since then, Argentina's economic policy has been schizophrenic, jerking wildly between radi- cally open markets and periods of intense control-freak economics, under which the country is currently suffering.

One time I was looking at a heat map showing product sales with an executive at a large European company. I noticed that there were high levels of sales in the countries surrounding Argentina but not much in Argentina itself. When I asked why, he told me that he would not invest in Argentina because it was nearly impossible to get capital out of the country due to government regulations restricting capital flows. He reported that the best way for non-Argentine companies doing business there to get their money out of the country is to buy huge quantities of beef that they then export abroad in exchange for dollars or euros. To this European executive, the hassle was not worth it. "I don't want to be in the beef business," he told me. Argentina's closed system has led to beef becoming a medium of international exchange—the polar opposite of what you'd hope for in a world where money is increasingly code-ified.

The tension between open and closed models is playing out with the highest level of tension in Muslim-majority countries around the world. To this point, oil-rich Muslim-majority countries like Saudi Arabia and Kuwait have gotten away with having closed societies but high GDPs because of oil wealth. With oil reserves set to dry up throughout much of the Gulf—some estimates assert that Saudi Arabia is likely to not be able to export oil in 15 years—these countries will no longer be able to rely on fossil fuel production in order to maintain their wealth and will have to pivot to knowledge-based industries.

The responses to this point have varied in terms of how far a state is willing to go. In Saudi Arabia, the government recently founded King Abdullah University of Science and Technology (KAUST). With a rumored investment of $20 billion, KAUST has sprung up from the dust of a remote fishing village. The gleaming new complex has attracted some of the world's leading researchers, prompting King Abdullah to dub it a new "House of Wisdom."

KAUST operates differently from research universities in open societies, often to its detriment. Visitors need advance authorization to visit the campus. Researchers must submit all proposals and publications for approval in order to guarantee that the work is "in the interest of Saudi Arabia." Scientific inquiry is designed for the pursuit of national interests, most notably fuel efficiency. KAUST faculty have fumed that it is merely an "important research lab for Aramco with a university façade."

In terms of social progress, KAUST merits some credit. Set against the background of Saudi Arabia's gender-restrictive culture, KAUST's openness to female faculty and students is significant. However, without those rights extending off-campus into Saudi society more broadly, it will be home to neither a headquarters nor a meaningful part of the supply chain for the industries of the future. The country as a whole remains too closed to make an impact, so it had better hope the oil continues to flow.

HALF OF THE WORLD

Going forward, a crucial factor in countries' success will be their ability to empower their own citizens—and this means *all* of their citizens. Too often countries still focus on only half—their male population—and ignore or abuse women, even as they hold so much potential.

The day after I met Maria Umar, just 200 kilometers away 15-year-old Malala Yousafzai was shot twice in the head and neck at point-blank range "for promoting Western thinking," according to a Taliban spokesman, and "propagating against the soldiers of Allah," according to the hooded man who shot her. What was she actually doing? Nothing more than asking that girls be offered an education, which the Taliban believes should be forbidden.

Hours after I first Maria Umar, I was at an event focused on entrepreneurship in Pakistan. I was at the head of a table lined with businessmen, half of whom had degrees from MIT. Sure enough, the event host sitting beside me said what I had heard everywhere else in the world: "We want to create our own Silicon Valley." I was the diplomat I was supposed to be and did not say that it was impossible. I spoke about the various attributes needed to foster innovation, entrepreneurship, and growth, and I complimented the Pakistani entrepreneurs for being the successful businessmen that they were.

Had it been a day later, after Malala had been shot, I probably would have said what I should have said anyway: "Forget it. As long as this is a country where fifteen-year-old girls get shot in the face and where schools are burned to the ground, the outlook here will be as bleak as the landscape in Waziristan."

It does not matter how many men have degrees from MIT if 90 percent of Pakistani women are victims of domestic violence and only 40 percent are literate. The states and societies that do the most for women are those that will be best positioned to compete and succeed in the industries of the future.

Treating women well is not just the right thing to do; it makes economic sense. Women are half of every nation's workforce—or potential workforce. To be a prosperous and competitive country requires access to the best-educated pool of workers. If a country is cutting off half of its potential workforce, it is taking itself out of the game. Countries that are closing the gender gap are competitive; they are the nations of the future, educating boys and girls and ensuring that their entire citizenry is skilled and ready for the global economy.

Put simply, nations that empower women reap the benefits. In the developing world, women can tip the scales between economic success and failure. Certain states are tricking themselves if they think they can compete and succeed without empowering women. Pakistan is the perfect example. Maria's and Malala's home regions are medieval. In Waziristan and the Swat Valley, you walk down dirt paths and men are pulling donkeys, railing against the West while their wives are locked up at home. Its society has created conditions so that the Maria Umars are too few and the victims of gender violence like Malala are too many.

According to the World Bank, 93 percent of Middle Eastern and North African countries have restrictions on the kinds of jobs that women can have. As long as these regions remain regressive, most American and European investors and executives will avoid them and pivot toward sub-Saharan Africa, Asia, and promising parts of Latin America.

This does not have to be the case in Muslim-majority countries. In Indonesia, I saw what has become one of the most intriguing and fastest-changing economies in the world. The world's largest Muslim-majority country with an estimated 250 million people, Indonesia is a sprawling mass of more than 17,000 islands that expand across more than 3,000 miles, farther than the distance from Seattle to Miami.

During my time in Indonesia I kept meeting entrepreneurs in

their 20s with 50 to 75 employees who had built their companies entirely cash-flow positive because they were not able to access venture capital or debt financing. If they did not make money, they could not pay people. It was as simple as that. Given that constraint, they nevertheless built a vibrant gaming and e-commerce community. In the offices of their companies, female coders sit next to male coders. Some wear hijabs, some don't. There is a subculture in Indonesia of the girl geek that in no way undermines their religious faith or practices. This culture extends to government too, where a law now mandates that at least 30 percent of a political party's candidates be women.

Ultimately the difference between Indonesia and countries like Pakistan and Saudi Arabia is about how a society interprets and applies religion. Indonesia has chosen to embrace Islam but not impose misogynistic laws. Pakistan has gone with an interpretation that too often keeps women at home and subjects them to beatings for demanding education. Meanwhile, many Gulf states are awkwardly claiming to be open while taking only baby steps in that direction. The best chance at producing the next home for the industries of the future among these nations? Indonesia.

WHAT CHINA AND JAPAN TELL US

The role of women in business and society was one of the most important and least-acknowledged drivers of the last stage of globalization, and their role will be even greater during the next stage. Women's equality is an issue not just in Muslim or developing countries but everywhere around the world, even in an advanced economy like Japan.

A look at the contrasting roles that women play in Chinese and Japanese business gives an indication of the benefits of women's empowerment in the workplace and the cost of their marginalization.

"Women hold up half the sky," Mao once said, and women's equality was central to his regime's efforts. The government urged women to open up small businesses in their neighborhoods and from their homes. In factories they were given nearly equal wages to men as well as child care benefits and flexible schedules. The progress of women in Chinese society over the course of decades is one of the major reasons it is the economic power it is today. A quarter of urban women go to college, where they outperform their male counterparts. In 2013, China led the world in the percentage of women in senior management positions—51 percent. Half of the world's wealthiest female billionaires live in China.

Jack Ma, Alibaba founder and CEO, told me over dinner one night that women were indispensable to Alibaba's success as both customers and leaders inside the company. One-third of its board is female, and nearly a quarter of top roles—vice president or higher—are filled by women. The ratio is still not 50-50, but it is substantially better than at the vast majority of technology companies. Ma told those of us at dinner that "the safest thing we do is lend to women entrepreneurs because they always pay it back." Ma was making the argument for empowering women not because it was fair or just but because it was good business.

By contrast, the role of women in business in Japan has contributed to its stagnation. The math suggests that it should not be that way. Japanese women are the best educated in the world. This high-quality education starts young; among OECD countries, Japanese girls score the best on standardized tests. In the years after college, though, Japan hemorrhages its female workforce. After women have their first child, 70 percent stop working for at least a decade, and many don't ever return to the workforce. (As a point of comparison, only 30 percent of American women do the same, and in Norway, 81 percent of mothers work.)

This fallout extends to every industry. Japanese women comprise less than 14 percent of university researchers and 19 percent of doctors. The numbers are no better in government, with Japan ranking 123rd out of 189 countries in gender diversity. And across the board, the World Economic Forum's Global Gender Gap Report in 2014 ranked Japan, one of the wealthiest nations in the world, 104 out of 142 assessed countries.

Given these statistics, it is not surprising that Japanese women are largely absent from leadership roles. Among executive-level managers, only 1 percent are women. Among the barriers is the fact that older Japanese men, with their traditional views of women, continue to dominate positions of power. They think women are first and foremost caretakers, and it is these men who make the hiring and promotion decisions.

Another barrier is a work environment that makes it nearly impossible to combine parenting and employment. Among full-time workers, one-fifth of Japanese 20 to 40 years old work more than 60 hours a week. That translates to either working every weekend or adding four extra hours to every standard workday.

For those who can get past the long hours, the after-work drinking culture is another barrier. Japan is a place where one is expected to socialize with lots of alcohol after work; it's an indispensable part of upward mobility in the Japanese workforce, and it almost always excludes female work colleagues. Basically, if you are a mother (or father, for that matter) who wants to see your kids, this work culture works against you.

China's rise and Japan's stagnation have been an embarrassment to the Japanese. With some coaching from Hillary Clinton, Prime Minister Shinzo Abe has begun to try to change this. At the core of his Abenomics economic plan, implemented after his election in December 2012, was a new place for women in the Japanese economy. In a speech at the World Economic Forum in Davos, Abe declared, "Japan must become a place where women shine." To make this happen, he has focused on increasing the availability of after-school programs for 10,000 kids. Too often, child care centers have lengthy waiting lists, so Abe has pushed for more private companies to open institutes. By contrast, China often relies on grandparents to serve as caregivers during the workday, doing so for 90 percent of the young children in Shanghai, 70 percent in Beijing, and half in Guangzhou.

eBay's former CEO John Donahoe has noted the importance of caregiving in his global workforce. He told me, "One of the most interesting things to me is that when I go to India or when I go to some parts of Asia, I notice there are so many women in our office. Why? It's because the grandparents are raising the children. Those societies are developing really interesting models. They are societies where young parents often live with their parents, or live next to their parents. Caregiving has adapted to the economy. The norms are that in their twenties and thirties, the young parents work to earn a living, and in their fifties or sixties they raise the grandchildren."

The economic importance of this fact pushed Prime Minister Abe to begin working on a modification to Japan's tax and pension policies to stop favoring stay-at-home wives rather than working women. He declared that by 2020, he wants women to be in 30 percent of Japan's leadership roles. "Japan's GDP could grow by 16 percent more, if women participated in labor as much as men," Abe said at the World Economic Forum in Davos. "That is what Hillary Clinton told me. I was greatly encouraged."

DIGITAL NATIVES

A second major condition necessary for societies to compete and succeed in the industries of the future is to have young people whose ideas are funded and whose place on organizational charts belies their youth. This will seem ridiculously obvious to anybody who has worked in Silicon Valley. It is less obvious to most others. At age 43, I am frequently the oldest person in any business meeting in Silicon Valley. In Europe, I am frequently the youngest.

While I don't believe that age automatically determines someone's proficiency with technology or aptitude to commercialize the businesses of the future, I do think that it matters whether somebody grew up living a digital life. Those who have see the world differently from someone like me who did not send an email or own a mobile phone until years after college. Digital natives are often less wedded to existing ways of doing business, and they are far more willing to take the kinds of risks that produce breakthrough innovations.

An early proof point for me was the Obama campaign in 2007 and 2008, when I served as the convener for technology, media, and telecommunications policy for the campaign. I was 35 years old at the beginning of the campaign. The campaign's chief technology officer, the chief digital officer, the person who ran analytics, the person who ran email, the person who ran social media: each of these people was younger than I was. Dan Wagner was just 24 years old when he ran national Get Out the Vote targeting efforts in that campaign, and four years later he was the chief analytics officer for Obama's reelection. Now he's in his early thirties and running a successful, fast-growing analytics company.

Dan draws a straight line between youth and an affinity and aptitude for analytics. Going through the profiles of 78 members of his staff, I only found one person with so much as a fleck of gray hair. This made me question whether Dan had it totally right, but it's hard to argue with his success, especially when I have seen plenty of its opposite in Europe.

I am utterly convinced that one of the unspoken reasons for France and Mediterranean Europe's prolonged stagnation is the degree to which young professionals are forced to wait for decades before being given real authority or the early-stage investments necessary to start their own companies. It is not a coincidence that Google, Facebook, Microsoft, Oracle, and countless other information-age companies were started by people in their twenties—and started in the United States.

As Dan Wagner says, "I think the United States is very special in the sense that in our culture we have an appreciation for merit and the best idea in the room. And it doesn't matter where you were born, where you're from, the color of your skin, your age. If you have a really good idea and you present that idea reasonably and thoughtfully to a group of people, they should accept that idea, and then employ that idea as the standard. But that's not always the case, right? But I think what people are increasingly realizing is if they don't integrate those ideas into what they're doing, they could be seriously in trouble, because somebody else will do it."

Venture capitalists in Silicon Valley won't hesitate to invest in twentysomethings. After leaving government, I began advising eight fast-growing companies that had financial backing from serious venture capitalists. At the time I started helping them, five of the eight had CEOs in their twenties, one had a CEO in his thirties, and two had CEOs in their forties. This would never have happened in a country like Italy, where someone in his or her twenties or thirties likely would have struggled just to be granted a short meeting with a venture capitalist and would then never have been trusted to run the company. It's a complaint I hear every time I travel to Spain, France, or Italy. Increasingly, if a young aspiring entrepreneur is not willing to wait until his forties to be taken seriously, he leaves and starts his company in a more youth-friendly culture like London, Berlin, or Silicon Valley.

Many Asian societies have also recognized how youth can turbocharge innovation in their countries. The average age of a CEO of a company listed on the Shanghai Stock Exchange is 47. By contrast, in more rigidly hierarchical Japan, the average age of a CEO on the Tokyo Nikkei Index is 62. China's biggest social media company was started by a graduate student in his 20s. Its largest e-commerce company was started by someone in his 30s, and the founder of its largest mobile phone company had just turned 40 and had already started several companies and invested in 20 more.

This is what it takes to compete now in the global economy. Dan asks, "What are you going to do in a world where you have young people coming in who have demonstrable merit that's greater than some of your executives who have been there for thirty years? Do you embrace those people, or do you chase them away?"

AFRICA: THE GREAT LEAPFROG OR PERMANENTLY LEFT BEHIND?

Can Africa—in its own way—pull off what India and China did in the last wave of globalization and innovation? With the fastest-growing population in the world and a strong talent base, African nations may be able to use the industries of the future to leapfrog ahead in development and even mitigate many of the costs that China and India faced during their breakout phase.

With 54 sovereign countries that make it as diverse as any continent on earth, Africa is difficult to characterize with any single sweeping statement. There are some trends, though, that apply nearly universally and that make a case for more optimism than pessimism when gauging how its countries will measure up in the industries of the future.

As I've traveled through Africa, I have continued to see an increasing number of examples of frugal innovation. In an environment of scarcity, people can become wildly creative. You could see this in Estonia after its independence. With no telecommunications or government infrastructure in place, it built systems with very few resources that were wildly creative and effective. I saw the same thing in Brazil, where this concept is known as *gambiarra*, and in rural India, where it is known as *Jugaad* innovation, for the Hindi word meaning "an innovative fix born from ingenuity and cleverness."

The creation of a product like M-Pesa is an example of frugal innovation at its finest. M-Pesa would only have been developed in a place like Kenya, a country without traditional banks serving the everyday needs of working-class people. In response, the Kenyans created an entire banking system using mobile phones and scratch cards. Thus, by frugal innovation, the country leapfrogged over the creation of a traditional banking system, at least as it exists in much of the rest of the world.

While many of the world's economies have stagnated since the economic crisis in 2008, Africa's have continued to grow at a fast pace. With this growth, Africans are increasingly both the founder-entrepreneurs as well as a part of global supply chains. More and more of Africa's technology-savvy youth are entering the workforce and starting their own companies or working remotely for Asian, American, or European companies. This is changing the nature of Africa's relationship with the rest of the world as its connections shift from being rooted in philanthropy and development aid to being rooted in business.

Jeremy Johnson is one of America's brightest young entrepreneurs. After launching two successful education companies before the age of 30, he founded Andela, a company to help connect Africa's rising technology geniuses to top employers. It helps Africa's rising stars, but there is nothing philanthropic about Jeremy's venture; it's a rewarding investment for all involved.

Andela has started work in Nigeria, where it launched job placement programs for the country's rising stars. In the company's first six months of operation, 9,597 young Nigerian professionals with an average age of 25 competed to enter Andela's boot camps. The admissions process is much more difficult than getting into Jeremy's alma mater, Princeton. In order to gain admission, Jeremy designed a test that he could not pass himself. One hundred seventy-eight Nigerians were admitted to the boot camps (30 percent of them women) and began the process of becoming Andela Fellows. The fellowship entails about six months of intensive training in coding and works toward placing fellows in technology jobs. At the first 12 companies that hired Andela Fellows, there was 100 percent retention, and 9 of the companies almost immediately asked for more. Jeremy says proudly that "our developers are some of the brightest, hardest-working young people not just in Africa, but anywhere."

The premise behind Andela was that there weren't just a few

genius technologists in Africa who needed some training and access to employers; there were thousands, maybe tens of thousands. All Andela developers have at least 1,000 hours of coding experience and can more than hold their own on any developer team.

Jeremy adds, "I expect them not only to hold senior positions at some of the most successful tech companies on the continent—I expect them to launch those companies one day. Creating opportunities for individuals is only the first stage—the real promise is empowering individuals to leave a lasting impact on their own communities and countries."

I saw the same level of technology genius on the other side of the continent in East Africa. In Tanzania, an East African nation of 45 million people, agriculture is essential for economic well-being. Grain is nicknamed "white oil" because it is so critical to economic growth. Agriculture provides 85 percent of exports and employs 80 percent of the workforce. The economy as a whole has often lurched back and forth with the vicissitudes of the grain market.

To stabilize the market and the wider economy, a 29-year-old Tanzanian computer programmer, Eric Mutta, developed an app called Grainy Bunch. It is a big data tool that uses apps to monitor the purchase, storage, distribution, and consumption of grain across Tanzania. Grainy Bunch took an ancient supply chain—grow grain, hope it thrives, sell it—and brought it into the 21st century. Analytics are now being used to better manage a valuable resource and improve access to food and returns for farmers. The effect has been to stabilize the grain market and help stabilize the larger Tanzanian economy.

In Kenya I was struck by the example of iCow, a text message– and voice-based mobile app being used by more than 11,000 small-scale dairy farmers. iCow, developed by a woman named Su Kahumbu, provides information for three stages of dairy farming: "menstruation, milking, and marketplace." In practice, this means that the app texts farmers on days of the cows' gestation periods, collects the farmers' milk and breeding records, and sends texts about dairy best practices.

One farmer joked to Kahumbu that "iCow tells me when to give my cow maternity leave."

iCow also alerts farmers to the days of highest demand for milk, veterinary information, and market price information. This last part basically means that instead of a farmer walking for half a day with his cow to the market and selling milk to whomever is standing in the town square at whatever price he is willing to pay, the farmer is now connecting to hundreds of possible buyers in the region through a mobile-enabled marketplace. If a farmer needs a veterinarian, she or he sends an SMS to iCow's short code with the word "VET," and iCow responds with the phone numbers of nearby veterinarians.

The average farmer using iCow owns just three cows. After seven months using iCow, the increased production is the equivalent of owning a fourth cow. For every dollar spent on iCow, the average farmer made an additional $77.

Both Grainy Bunch and iCow were part of the Apps4Africa program that we launched during my time at the State Department. Apps4Africa matches innovative African tech start-ups with the cash to get their businesses going and takes advantage of what is now more than 650 million mobile phone subscriptions on the African continent, more than in Europe or America.

In addition to having the technology expertise, Grainy Bunch and iCow both reinforce the theory that wherever there is domain expertise and a willingness to apply big data technologies, there is an opportunity to create the businesses of the future. There are huge supply-chain management software companies in California and Germany, but Grainy Bunch was developed in a place with deep understanding about the supply chain for grain and grain markets. iCow was developed specifically for low-literacy dairy farmers who own just a few cows, the complete opposite of New Zealand, where Pasture Meter was developed and dairy herds frequently number in the thousands.

Su Kahumbu is also part of a larger trend in sub-Saharan Africa, which (along with Latin America) has the highest rates of gender parity in entrepreneurship in the world. Many nations in Africa are benefiting from an increased role for women and youth in their economies. And the recent increase in women's economic roles in the African economy has corresponded to the continent's longest and largest period of economic growth. "Women in the private sector represent a powerful source of economic growth and opportunity," said Marcelo Giugale, the World Bank's director for poverty reduction and economic management for Africa. And indeed, women have been central to Africa's rapid, leapfrogging growth. In a number of countries the rate of female entrepreneurship is equal to that of men, and Nigeria and Ghana (representing about 25 percent of sub-Saharan Africa's population) in fact have more female than male entrepreneurs.

Perhaps the most striking example I saw of a nation using technology to leapfrog economically was Rwanda. Two decades after the brutal genocide of 1994, which saw more than 800,000 people murdered, Rwanda has reimagined and rebuilt itself with a knowledge-based economy at its core.

No border crossing has been more memorable for me than that between the Democratic Republic of Congo and Rwanda. At the Congo's eastern border, there is chaos. The queue is hours long, men with guns shake people down for bribes, and the roads look bombed out. When you cross the border from the Democratic Republic of Congo into Rwanda, you notice that the roads are suddenly smooth and well paved. The worst road in Rwanda is better than the best road in the Congo. Driving east, while I'm still in the jungle, my smartphone chirps to life with five bars of data-enabled connectivity.

Driving through the hills of western Rwanda toward the capital city of Kigali, I regularly passed shoulder-high spools of fiber lining the road to lay a fiber-optic network better than those in much of the rural United States. It now connects all 30 of Rwanda's districts with 1,000 miles of fiber, allowing a little country in the center of Africa to connect to the wider world and open a high-tech commodities exchange.

If you look at the math, the strategy has worked. Between 2001 and 2013, real GDP growth averaged over 8 percent a year and poverty decreased substantially. Unlike many other economies (including the United States) where inequality has increased despite overall economic growth, Rwanda's inequality has decreased over the past 15 years.

Though not a favorite of journalists and (some) human rights advocates, Rwanda's president, Paul Kagame, has taken a landlocked African country that exhibited the most savage behavior humanity is capable of and turned it into a nation with a functional economy and an innovation strategy at its core. The idea is for Rwanda to move straight from an agricultural economy to a knowledge-based economy, bypassing the industrial phase altogether.

The results have been promising, and Kagame attributes much of the success so far to systematic efforts toward reducing all barriers to participation for women. His government made gender equality a key tenet during its postconflict reconstruction, instituting policy and legal reforms that ensured equal rights for women and banned gender-based violence.

In a nation long reliant on agriculture, a policy shift for land titles allowed wives to be registered alongside their husbands. This move proved critical: Rwanda saw a 20 percent increase in female-registered farms. Meanwhile the rates of Rwandan women in poverty dropped by nearly 20 percent.

When I spoke with Kagame at his modest home in Kigali, he walked through his strategy to help Rwanda become a knowledge-based economy. As I asked about the role of women in tomorrow's economy, he teased me a little, noting that women in Rwanda constituted a larger percentage of leaders in the public and private sector in Rwanda than in the United States. It turns out that Rwanda is the *only* country in the world with a democratically elected parliamentary body that is majority female.

If Rwanda can go from the trauma of genocide to a growing, diverse, knowledge-based economy, then it can happen anywhere.

Andela's Jeremy Johnson says that "Africa is witnessing the convergence of demographic, economic, and technological trends that hold incredible promise for its future. The combination of a young population, fast-growing economies, and rapid technology adoption is creating a dynamic engine for private sector investment."

What I have seen in Africa makes me believe that industries of the future will have more broadly distributed centers of innovation and wealth creation than was the case in the past 20 years, when Silicon Valley dominated all comers. The businesses being built in Africa make smart use of big data without being reliant on platforms built in Silicon Valley. These solutions give me hope that big data will allow more businesses to innovate wherever they are, in effect creating more opportunities in more places around the world than has ever been possible before now.

Africa also validates my belief that the societies that embrace openness will be those that compete and succeed most effectively in the decades to come. Many African countries are still far from being as politically open as they could or should be; but those that have opened up economically, those that have empowered the women in their society, and those that have created space for their entrepreneurs are growing the fastest. There has never been a better time to do business in Africa. And just as India worked its way out of its caricature as the country of Mother Teresa, famine, and a nasty caste system, so too are the countries of Africa recasting themselves in the world as being a place for investment instead of assistance.

What is true in Africa is true for the rest of the world. When leaders wonder what they can do to position their societies for the industries of the future, they need to open up and resist control-freak tendencies. The 21st century is a terrible time to be a control freak; future growth depends on empowering people.

CONCLUSION: THE MOST IMPORTANT JOB YOU WILL EVER HAVE

Robots that care for us as we grow old. Cyberattacks against our homes. Extinct animals brought back to life. Ubiquitous sensors eliminating privacy as we now know it. These changes are disorienting and more than a little scary. As much as I think about the changes I have described in this book from an economic and geopolitical perspective, what really gets popcorn popping in my head is thinking about these changes from my perspective as the father of three children ages 13, 11, and 9.

The most important job I will ever have is being a dad, and I can't help wondering what all these coming changes—the ones that this book anticipates and the ones that it does not—will mean for our children's economic future. My kids will have an entirely different set of opportunities and challenges than I had growing up in West Virginia. What will it take for them to compete and succeed?

I asked just about everyone I interviewed for this book what attributes today's kids will need for tomorrow's economy. There was no consensus—no single, singular conclusion to put in a headline. But there was near-consensus on a thing or two and some common themes that emerged as I talked to more people.

To start, the stories of the two youngest people interviewed for this book give good glimpses into the attributes that today's children will need for tomorrow's economy.

Think back to 24-year-old venture capitalist Sheel Tyle, who set out on his career path after being inspired by Sudanese mobile phone billionaire Mo Ibrahim. Sheel's parents are both from India and came to the United States for higher education. His mother, Tanu (the first in her family to fly on an airplane), was one of 15 women in a class of 1,000 at her college in India, which prompted her to move her studies to the United States. Sheel's father, Praveen, applied to colleges that did not charge an application fee. He went to Ohio State University instead of other universities, including the Ivy League, because Ohio State gave him a full scholarship and he received a free plane ticket to travel there.

As Sheel's parents entered the professional class in America, they decided to take Sheel and his younger brother, Sujay, on trips that would help them understand that they were living lives of relative privilege, contributing to their emotional development and making them more worldly. Sheel says that "when we were growing up, we never took trips to Europe or the Caribbean. Anytime my parents had some free time, they wanted to show us how the real world works."

His parents took them to Brazil and Kenya in the 1990s when both were still considered undeveloped frontier countries. When he was seven years old, Sheel's family traveled to an orphanage for blind children, 80 percent of whom had treatable blindness but could not be treated because of the lack of funds.

His parents were not wealthy, but they spent a big piece of their incomes on these travels to open up their kids' eyes to the wider world. Sheel and his brother were little kids, but they were already imagining their lives and careers being played out in a global context. That was why Mo Ibrahim's success in bringing mobile telecom to Africa set Sheel on his course as an investor.

In the same way that entrepreneurs, businesses, and investors who engaged in China and India ten to twenty years ago were able to build big businesses, people who can look around the world and see and understand the opportunity in the next wave of high-growth markets are those who will realize the greatest gains. The time Sheel spent in places like Nairobi was seminal for him, and as he invests in the hottest of hotshot early-stage deals in Silicon Valley—in fields including cryptocurrency, clean tech, consumer Internet, and mobile—he's also doing something that only a tiny percentage of Silicon Valley investors are doing: investing in places that are today's frontier markets, like Kenya, Uganda, and Bangladesh. As those markets develop, those like Sheel who are educated about the market will have a head start toward developing the relationships and partnerships to source quality investments. They will get in early, when valuations are at their lowest—where China stood in the 1990s and the Internet stood in 1994.

Sheel imagines himself always working globally, and he thinks of that huge geographic span as home. He says, "I don't aspire to nor do I feel like our circle of friends will truly settle anywhere; we are constantly going between the San Francisco–Boston–New York–DC corridor domestically and then large emerging market cities. Home for me isn't a place but rather a feeling—a feeling best felt when near family or with close friends."

Today Sheel is the youngest venture capitalist with a senior role at a major Silicon Valley venture capital firm. His brother, Sujay, matriculated at Harvard as a 15-year-old and stayed for five semesters before accepting a Thiel Fellowship. The fellowship, set up by PayPal Mafia alum Peter Thiel, gives young college students $100,000 to drop out of college and focus on entrepreneurship. Sujay moved out west and became the chief operating officer of Hired.com (an online marketplace where companies compete for engineering talent) and a vice president at a mobile entertainment network. He recently decided to go back to school to finish up an environmental science and public policy degree at Harvard.

Ten years older than Sheel is Jared Cohen, and at just 34, he is still young in my eyes. When I first went to work for Hillary Clinton at the State Department at the beginning of Obama's presidency, I met Jared, then 27 years old. He was one of the few holdovers from the Bush presidency. By the time I met him, he was a Rhodes scholar and had already written two books. Like Sheel, he had gone to Stanford for his undergraduate education. Jared and I worked closely together for a year and a half before he left to work for Google chairman Eric Schmidt and establish Google Ideas. My experience traveling and working with Jared reinforces what I think can be learned from Sheel.

The son of a psychologist and an artist in Connecticut, Jared grew up with a curiosity about foreign languages and cultures. He began teaching himself Swahili from a book when he was 16 years old, a sophomore in high school. His mom then began taking him to private Swahili classes at Yale, and he began traveling to Africa. At 19, he lived with the Masai tribesmen in Kenya.

While Jared and I were together in East Congo and the hills in western Rwanda, it was no small advantage to have a fluent Swahili speaker on the team. We were able to bypass the Abbot-and-Costello-like translation plan the embassy set up, where the locals would speak Swahili to an African translator, who would translate it to French for a local embassy staffer, who would then translate from French to English for me and Jared. Instead we were able to communicate and engage directly with people, from militia members who were being repatriated to Rwanda to victims of sexual violence in the refugee camps in East Congo.

Our ability to develop successful programs in the region took advantage of the fact that we were fluent in both the technology and the local language and culture. It's that same dynamic that allowed Mo Ibrahim, the Sudanese mobile phone billionaire, to build businesses in frontier markets including the Congo. It's the willingness and ability to immerse yourself in today's frontiers that will create many of tomorrow's big businesses. And it is people like Sheel and Jared who will see the opportunities first and will have the skills and relationships to take advantage of those opportunities. Ironically, in a world growing more virtual, it has never been more important to get as many ink stamps in your passport as possible.

Most people can't afford family travel to frontier markets like Sheel's family or private Swahili lessons at Yale like Jared's, but today's parents have many tools that did not exist as recently as Sheel and Jared's childhoods. Language-learning programs are available online that are nearly as good as what can be gotten from a private tutor. There is no substitute for getting on a plane and traveling to frontier markets to learn about them, but the choices made by Sheel's and Jared's solidly middle-class parents put them on a course to achieve the steep upward economic and social mobility they enjoy today.

If a major lesson learned from Jared and Sheel is that multicultural fluency is increasingly important in a business world that is growing more global, other thinkers and experts I spoke to emphasized a different set of skills—or said that foreign language skills were only part of the equation. Many believe that today's kids must also become fluent in a technical, programming, or scientific language. If big data, genomics, cyber, and robotics are among the high-growth industries of the future, then the people who will make their livings in those industries need to be fluent in the coding languages behind them.

"If I were eighteen right now, I would major in computer science or engineering, and I'd be taking Mandarin," former eBay CEO John Donahoe told me. He used his son as an example of what he believes is the right approach: "My youngest son's a freshman at Dartmouth. He's taken Mandarin for four years and he'll probably major in computer science."

Investor and entrepreneur Chamath Palihapitiya shared with me the approach that he and his wife, Brigette Lau, also a computer engineer, bring to parenting their two children: "I think it's really important that people have at least two other languages: one that is traditionally classically linguistic and one that's technical. And the reason is because the way the human capital markets are changing, you need to have this solidity to be able to converse with people in different parts of the world, understand their cultures, understand their languages, as well as being able to converse technically. My approach in our family is my kids need to learn two languages; one is Spanish— they've learned it from day one—and the second will be like Python or some other technical language, which they'll learn when they're six and older. That's the one important thing that we've decided, that languages are going to be a really important way to facilitate an understanding of the world, both the physical world in which we live as well as the technical world in which we live."

The importance of learning technical languages comes up again and again. Charlie Songhurst provided an interesting counternarrative. He sees today's need for highly technical and mathematical skills as a short-term phenomenon. "There's a demand curve for certain skill sets at a given time," he says. "At the moment there's a demand for aspergy-math minds. But I think we've only got ten more years of the Asperger's economy, because once the tech platforms are established, they won't reinvent."

In contrast, Jack Dorsey makes the case that the benefits of programming language fluency go well beyond coding: "I don't think you do it to become an engineer or to become a programmer; you do it because it teaches you how to think in a very, very different way. It teaches you about abstraction around breaking problems into small parts and then solving them, around systems and how systems interconnect. So these are all tools you will use everywhere, especially as you think about building a business, or running a business, or even working in a business. If you can synthesize a massive, complex system into something that is essential that you can articulate in a very crisp way, that's exactly what programming teaches you."

Google's Eric Schmidt reinforces Jack's point about the importance of learning how to understand complex problems. When I asked Eric what skills he thought my kids would most need, he told me that "the biggest issue is simply the development of analytical skills. Most of the routine things people do will be done by computer, but people will manage the computers around them and the analytical skills will never go out of style."

For this reason, many of the people I spoke to encouraged the age-old liberal arts education and its credo of "learning how to think." Indeed, many felt that the distance between traditional liberal arts fields and engineering fields would begin to collapse. Jared Cohen asks, "Why should I have to be a political scientist or a computer scientist? Why is there not a hybrid between the two? Why is it that I have to be either a historian or an English major or an electrical engineer? Why is there no hybrid between the two? You know they are both languages. The point is there needs to be a more interdisciplinary approach that merges the sciences and the humanities in a way that prepares kids for a world where those silos are already beginning to be broken down."

Jared is making the point that today's parents should raise their children in the way that Sheel and Sujay Tyle's parents raised them, sending them off to university studies in human biology and public policy for Sheel and environmental science and public policy for Sujay.

Estonia's president Toomas Ilves makes a similar point, suggesting that domains previously occupied *only* by people with backgrounds in the liberal arts, like government, will become increasingly occupied by people with more background knowledge in science and technology. He points to the example of his technology-savvy son, Luukas, who works in government: "He's never going to invent a billion-dollar app, but he's in policy, and he understands the policy implications, and that is, I think, one of our problems right now: we don't have, in Europe at least, people at the policymaking level who understand what IT is about."

But what about the many children born around the world who will not have access to college? There are several resources that have arisen lately that democratize access to important programming skills. One is Codeacademy, a Y Combinator project cofounded by two 23-year-olds that teaches people how to code for free online. Codeacademy counts more than 24 million people around the globe who have used its resources. A second incredible resource is Scratch, a project of the Lifelong Kindergarten Group at the MIT Media Lab. It's a nonprofit endeavor that teaches programming. It is free and doesn't require a download. It is well-suited to low-bandwidth environments and available in more than 40 languages. To date, more than 5 million projects have been developed on Scratch in more than 150 countries, so it is just about everywhere.

Today's youth who will enter tomorrow's workforce will need to be more nimble and more familiar with the broader workings of the world to be able to find a niche that they can fit into. With robotics automating labor that is cognitive and nonmanual, the kind of job that my father made a 50-year-long career of—practicing real estate law—would be a bad bet for someone exiting law school today. Tomorrow's labor market will be increasingly characterized by competition between humans and robots. In tomorrow's workplace, either the human is telling the robot what to do or the robot is telling the human what to do.

Children growing up in environments of economic and social privilege will always have an advantage over those growing up under lesser circumstances. Much of that privilege has been determined over the years by geography. Throughout the 20th century, the single greatest economic advantage one could have was to be born in the United States or Europe. That relative economic benefit—the spread between the United States or Europe and the rest of the world—has decreased over the last 20 years. As what were previously frontier markets like

China, India, Indonesia, and Brazil became fast-developing markets, there was substantial growth of those countries' middle classes and their elites. In addition to the billion people who entered their middle classes, there are now more than 200 billionaires in China, 90 in India, 50 in Brazil, and 20 in Indonesia.

Living in a fast-growing market provides a rare opportunity to achieve upward economic mobility. And just as China, India, Brazil, and Indonesia were among the past beneficiaries of this growth, we can now say that there has never been a better time to be born in sub-Saharan Africa, where once poor, isolated communities are increasingly part of the global economy and the source of what will be much of the next decade's growth. As more resources like Codeacademy and Scratch spread without geographic constraint and as more companies like Andela invest in today's frontier markets, the world will have more fast-developing economies. Best positioned to succeed will be the countries that open up economically, politically, and culturally.

The growing economic diversity and increasing pace of change means that investors and people in global business will have to be as mobile and able to work across cultures as people newly entering the workforce. The same advice that applies to the next generation applies to today's investors if they want to be a part of the trillions of dollars of wealth creation that will come from the industries of the future. The innovation and company creation that is just now beginning to take place in robotics, genomics, cyber, big data, and new fields made possible by the code-ification of money, markets, and trust will spring from alpha cities around the world, but they will also come from places that most business leaders have never visited, like Estonia.

The rise of the Internet economy has taught business leaders that very young people who grew up digital are likely the ones who are going to create the big Internet companies. The same will hold true in many of the industries of the future. I expect most of the billion-dollar businesses in cyber and big data to spring from the minds of people in their twenties and thirties—those who grew up programming in a time of code war and the exponential growth of data.

I often think back to the midnight shift on the janitorial crew. For many people I met on that job, their entire professional lifetimes would be spent pouring chemicals on the floor after a country music concert, even as they were capable of much more—if they'd simply had an option for career growth or the chance to go back to school.

There is no shame in these jobs, but there is great shame for society and its leaders when a life is made less than what it could be because of a lack of opportunity. The obligation of those in positions of power and privilege is to shape our policies to extend the opportunities that will come with the industries of the future to as many people as possible.

For most of the world's 7.2 billion people, innovation and globalization have created opportunity the likes of which has never before existed. The number of people who have recently moved out of poverty in China alone is equal to double the population of the entire United States. The number of people living in severe poverty and able to concern themselves only with meeting the basic needs of food, shelter, and clothing has decreased at a rate previously unknown in human history.

These changes mean new opportunities for all of us—for businesses, governments, investors, parents, students, and children. This book, I hope, will help us to make the most of them.

ACKNOWLEDGMENTS

The book begins with the story of my working the midnight shift as a janitor. I owe my parents, Alex and Becky Ross, a debt of gratitude for making me work those tough jobs. Those jobs made me who I am today.

Twenty years after the midnight shift ended for me, Hillary Clinton made a bet that we could launch an innovation agenda that would dramatically advance America's diplomatic and developmental goals. The four years I spent as her Senior Advisor gave me the privilege of public service and the insights that informed the writing of *The Industries of the Future*. Thank you, Madame Secretary.

I am profoundly grateful to Jonathan Karp and Jonathan Cox for their wisdom and diligence, taking what was just the beginning of an idea and working with me to turn it into this book. My first draft was a 200,000-word mess, the product of what could only be a first-time author. Their sustained attention, coaching, and copious editing produced the book you have in your hands now.

My gratitude goes to Ariel Ratner for applying his 200-point IQ to test every assumption, scrutinize every sentence, and never allow this book to lose its soul. He was an indispensable partner.

I deeply appreciate to the work of my agent, Greater Talent Network's Don Epstein, for knowing that I had this book in me, and then making it happen.

I relied on a group of essential friends and advisors who offered wisdom, judgment, and sympathy throughout the writing process. Among the many, I must single out Jared Cohen, Ari Wallach, Ben Scott, Jonathan Luff, and Robert Bole. Jared always offered encouragement and insight when I needed it most. He kept me ambitious. Ari understood the *why* of this book better than anybody. Everybody should have a rabbi like Ari, regardless of faith. Ben has the sharpest policy mind of anybody I have ever worked with. He works behind the scenes in constant service of the public good. He also protects me from myself, a thankless act. Jonathan and Rob indulge heavy doses of both bluster and blue funk from me. Thank you for your friendship.

I drew on the skills of a great many people to help turn ideas into thoroughly researched prose. My thanks to Teal Pennebaker for unraveling the mysteries of the genome's potential and for ensuring that the equities of women remained at the core of this work. Thank you to Olga Belogolova for exploring the dark, dangerous work of the cyberdomain.

Thanks to the numerous interns and researchers who labored long and hard. They include Jennifer Citak, Shana Mansbach, Alissa Orlando, Christopher Murphy, Tristram Thomas, Nimisha Jaiswal, Saraphin Dhanani, Fiona Erickson, Paul Mayer, and Kate Galvin.

Special thanks to the people who have been so supportive of me as I have taken on a life after government that is more than a little crazy. They include Ron Daniels and the faculty of Johns Hopkins University, where I currently serve as a Distinguished Visiting Fellow, and Merit Janow, Dan McIntyre, and the faculty of Columbia University's School of International and Public Affairs, where I served as a Senior Fellow for two academic years. Thank you to Marvin Ammori; Jose Andres; Matthew Barzun; Shawn Basak; Avish Bhama; Elana Berkowitz; Ian Bremmer; Grace Cassy; Farai Chideya; Scott Crouch; Bill DePaulo; Raymond DePaulo; Katie Dowd; Georgeta Dragoiu; Guy Fillippelli; Charlie Firestone; Alan Fleischmann; Julius Genachowski; David Gorodyansky; Julia Groeblacher; Alex Gurevitch; Craig Hatkoff; Reid Hoffman; Reid Hundt; Tim Hwang; Christian Johansson; Jeremy Johnson; Bettina Jordan; Jed Katz; Bill Kennard; Andre Kudelski; Eric Kuhn; Jeffrey Leeds; Blair Levin; Peter Levin; Jason Liebman; Catherine Lundy; Adam, Allison, Dave, and Robyn Messner; Bruce Mehlman; Yuri Milner; Wes Moore; Maryam Mujica; Craig Mullaney; Marc Nathanson; Colm O'Comartun; Chip Paucek; Andrew Rasiej; Wayne and Catherine Reynolds; Jane Rosenthal; Stephen Ross; Eric Schmidt; Joshua Stern; Mark Tough; Roman Tsunder; Sheel Tyle; and the indefatigable Rebecca Wainess.

NOTES

INTRODUCTION

4 *The Soviet Union and its satellite states*: "1991: Hardliners Stage Coup against Gorbachev," *BBC News*, On This Day, http://news.bbc.co.uk/onthisday/hi/dates/stories/august/19/newsid_2499000/2499453.stm; "Fall of the Soviet Union," History.com, http://www.history.com/topics/cold-war/fall-of-soviet-union.

4 *India began a series of economic reforms*: "India's Economic Reforms," India in Business, Ministry of External Affairs, Government of India, Investment and Technology Division, http://www.indiainbusiness.nic.in/economy/economic_reforms.htm.

4 *China reversed its economic model*: "Poverty & Equity Data | China," World Bank, http://povertydata.worldbank.org/poverty/country/CHN.

4 *The North American Free Trade Agreement (NAFTA) came into effect*: World Trade Organization International Trade Statistics, 2013, World Trade Organization, http://www.wto.org/english/res_e/statis_e/its2013_e/its2013_e.pdf.

5 *The highest-skilled labor markets*: Richard Rahn, "RAHN: Estonia, the Little Country That Could," *Washington Times*, June 20, 2011, http://www.washingtontimes.com/news/2011/jun/20/the-little-country-that-could/.

6 *A computer that can speed up analysis*: John Markoff, "Armies of Expensive Lawyers, Replaced by Cheaper Software," *New York Times*, March 4, 2011, http://www.nytimes.com/2011/03/05/science/05legal.html?pagewanted=all.

6 *Social networks can open doors*: Larry Rosen, *iDisorder: Understanding Our Obsession with Technology and Overcoming Its Hold on Us* (London: Palgrave Macmillan, 2012).

6 *The digitization of payments*: Luke Landes, "What Happens If Your Bank

Account Is Hacked?" *Forbes*, January 15, 2013, http://www.forbes.com/sites/ moneybuilder/2013/01/15/what-happens-if-your-bank-account-is-hacked/.

7   *The Union Carbide Corporation established*: "South Charleston Manufacturing Site," The Dow Chemical Company, West Virginia Operations, http:// www.dow.com/ucc/locations/westvir/awv/inf03.htm.

8   *Between 1946 and 1982*: "Union Carbide Corporation," *West Virginia Encyclopedia*, http://www.wvencyclopedia.org/articles/823.

8   *Between 1946 and 1982*: Ibid.

8   *By 1960, its population*: "Census of Population and Housing, 1960," US Census Bureau, http://www.census.gov/prod/www/decennial.html.

9   *For almost a century*: Laura Parker, "A Century of Controversy, Accidents in West Virginia's Chemical Valley in Lead-up to Spill," *National Geographic*, January 16, 2014, http://news.nationalgeographic.com/ news/2014/01/140116-chemical-valley-west-virginia-chemical-spill-coal/.

10   *The war ended*: "nitro, w.va.," WVCommerce.org, http://www.wvcommerce .org/people/communityprofiles/populationcenters/nitro/default.aspx.

10   *In the 1960s*: "Agent Orange: Background on Monsanto's Involvement," Monsanto Company, http://www.monsanto.com/newsviews/pages/agent-orange-background-monsanto-involvement.aspx.

10   *Dropping this chemical in the Vietnamese jungle*: "Agent Orange and Veterans: A 40-Year Wait," White House, http://www.whitehouse.gov/ blog/2010/08/30/agent-orange-and-veterans-a-40-year-wait.

10   *From 1960 to 1990*: "Census of Population and Housing, 1960" and "Census of Population and Housing, 1990," US Census Bureau, http://www.census .gov/prod/www/decennial.html.

10   *By 1988, West Virginia's unemployment rate*: "Unemployment Rate US, seasonally adjusted" (series ID: LNS14000000), US Department of Labor, Bureau of Labor Statistics, http://beta.bls.gov/dataViewer/view/time-series/LNS14000000; "Local Area Unemployment Statistics, West Virginia" (series ID: LASST540000000000003), US Department of Labor, Bureau of Labor Statistics, http://beta.bls.gov/dataViewer/view/timeseries/ LASST540000000000003.

11   *The steel mills of Pittsburgh*: Nick Carey, "Detroit Jobs Might Return, But Workers Still Lack Skills," Reuters, August 2, 2013, http://www.huffingtonpost .com/2013/08/02/detroit-jobs-_n_3693303.html.

11   *Manchester, the world's first industrialized city*: "Manchester/Liverpool," *Shrinking Cities*, http://www.shrinkingcities.com/manchester_ liverpool.0.html?&L=1; "Manchester—the First Industrial City," Science Museum, http://www.sciencemuseum.org.uk/on-line/energyhall/page84 .asp.

11   *The coalfields in southern Wales*: "Coal Mine Closes with Celebration," *BBC News*, January 25, 2008, http://news.bbc.co.uk/2/hi/uk_news/ wales/7200432.stm.

11   *The port of Marseilles was ravaged*: Olaf Merk and Claude Comtois,

"Competitiveness of Port-Cities: The Case of Marseille-Fos, France," OECD Library, http://www.oecd-ilibrary.org/docserver/download/5k8x9b92cnnv .pdf?expires=1404165171&id=id&accname=guest&checksum=50B32B0E0 157BCABC82720A0251D05E2.

11   *In the 30 years from 1982 to 2012*: "World Development Indicators," World Bank, http://data.worldbank.org/country/india.

11   *Life expectancy surged*: Ibid.

11   *The changes in China*: "Country Partnership Strategy for the People's Republic of China for the Period FY2013-FY2016," World Bank, International Bank for Reconstruction and Development, International Finance Corporation, and Multi-lateral Investment Guarantee Agency, Report 67566-CN, October 11, 2012, http://www-wds.worldbank.org/external/default/WDSContentServer/WDSP/IB/2012/ 11/12/000350881_20121112091335/Rendered/PDF/NonAsciiFileName0.pdf.

11   *With an economy 25 times larger*: "World's Largest Economies," *CNN Money*, http://money.cnn.com/news/economy/world_economies_gdp/.

CHAPTER 1: HERE COME THE ROBOTS

15   *Japan's current life expectancy*: "Population Projections for Japan (January 2012): 2011 to 2060," National Institute of Population and Social Security Research, January 2012, http://www.ipss.go.jp/site-ad/index_english/ esuikei/ppfj2012.pdf.

15   *Between 2010 and 2025*: "Japan Moving toward Nursing Robots for Elderly," *Japan Economic Newswire*, June 12, 2013, http://asq.org/qualitynews/qnt/ execute/displaySetup?newsID=16207.

15   *Today, 25 percent of Japan's*: "Population Ages 65 and Above (% of Total)," World Bank, http://data.worldbank.org/indicator/SP.POP.65UP.TO.ZS.

15   *By 2020, this is projected*: Japan Ministry of International Affairs and Communications, Statistics Bureau, *Statistical Handbook of Japan—2014*, http:// www.stat.go.jp/english/data/handbook/c02cont.htm; John Hofilena, "Japan Pushing for Low-Cost Nursing Home Robots to Care for Elderly," *Japan Daily Press*, April 29, 2013, http://japandailypress.com/japan-pushing-for-low-cost-nursing-home-robots-to-care-for-elderly-2927943/.

16   *Right now there are only 1.49 million*: "Difference Engine: The Caring Robot," *Economist*, May 24, 2013, http://www.economist.com/blogs/bab-bage/2013/05/automation-elderly.

16   *He can do the dishes*: "Partner Robot Family," Toyota: Innovation, http:// www.toyota-global.com/innovation/partner_robot/family_2.html.

17   *Equipped with cameras that function as eyes*: Lee Ann Obringer and Jonathan Strickland, "How ASIMO Works," *HowStuffWorks*, http://science .howstuffworks.com/asimo1.htm.

17   *For an elderly patient*: "Seven Robots That Can Help Aging Americans," *Fiscal Times*, May 2, 2013, http://www.thefiscaltimes.com/Media/ Slideshow/2013/05/02/7-Robots-That-Help-Aging-Americans.aspx?index=2 #zFQXE8DZODxK7z2p.99.

17   *Its Walking Assist device*: "Walking Assist: Supporting People with Weakened Leg Muscles to Walk," Honda—Products and Technology, http://world .honda.com/Walking-Assist/.

17   *Numerous other Japanese companies are pushing*: "World's First Robot That Can Lift Up a Human in Its Arms," RIKEN-TRI Collaboration Center for Human-Interactive Robot Research, http://rtc.nagoya.riken.jp/RIBA/ index-e.html; Grace Liao, "Meet RIBA-II, RIKEN's New Care-Giving Robot for Japan's Elderly," *Asian Scientist Magazine,* August 3, 2011, http://www .asianscientist.com/in-the-lab/meet-riba-ii-rikens-care-giving-robot-japans- elderly/.

17   *Designed for those who are too frail*: Michael Fitzpatrick, "No, Robot: Japan's Elderly Fail to Welcome Their Robot Overlords," *BBC News Tokyo,* February 4, 2011, http://www.bbc.co.uk/news/business-12347219.

17   *When President Barack Obama met PARO*: "Obama Test Drives Japanese Technology," YouTube, https://www.youtube.com/watch?v=CfCTBOTHsVU.

17   *It looks like a cute stuffed animal*: Anne Tergesen and Miho Inada, "It's Not a Stuffed Animal, It's a $6,000 Medical Device," *Wall Street Journal,* June 21, 2010, http://online.wsj.com/news/articles/SB10001424052748704463504575301051844937276?KEYWORDS=paro&mg=reno64-wsj&url=http%3A%2F%2Fonline .wsj.com%2Farticle%2FSB10001424052748704463504575301051844937276 .html%3FKEYWORDS%3Dparo.

18   *In 2013, the Japanese government granted*: "Will Your Golden Years Be Robot-Assisted?" *Techonomy,* May 6, 2013, http://techonomy.com/2013/05/ will-your-golden-years-be-robot-assisted/.

18   *Japan's prominent Ministry of Economy, Trade, and Industry*: "Japan Moving Toward Nursing Robots for Elderly," *Japan Economic Newswire,* June 12, 2013, http://asq.org/qualitynews/qnt/execute/displaySetup?newsID=16207.

18   *Tasks for these robots include*: Ibid.

18   *On the technical side, it remains*: "Difference Engine: The Caring Robot," *Economist,* May 14, 2013.

18   *We are building the machines*: Clara Moskowitz, "Human-Robot Relations: Why We Should Worry," *Live Science,* February 18, 2013, http://www .livescience.com/27204-human-robot-relationships-.html.

19   *In Europe, all 28 member states*: Eurostat, European Commission, "Population Structure and Ageing," http://epp.eurostat.ec.europa.eu/statistics_explained/ index.php/Population_structure_and_ageing; "European Commission Ageing Report: Europe Needs to Prepare for Growing Older," May 15, 2015, http:// ec.europa.eu/economy_finance/articles/structural_reforms/2012-05-15_ageing_ report_en.htm.

20   *A few countries have already established*: "Industrial Robot Statistics: World Robotics 2014 Industrial Robots," International Federation of Robotics, http://www.ifr.org/industrial-robots/statistics/.

20   *Japan, the United States, and Germany dominate the landscape*: Josh Bond, "Robot Report Predicts Significant Growth in Coming Decade," *Logistics*

Management, April 25, 2013, http://www.logisticsmgmt.com/article/robot_report_predicts_significant_growth_in_coming_decade.

21 *One winner was RoboArm*: Nathan Hurst, "These $10 Robots Will Change Robotics Education," *Wired*, September 29, 2012, http://www.wired.com/2012/09/afron-winners.

21 *As a result, Japanese culture tends*: Christopher Mims, "Why Japanese Love Robots (and Americans Fear Them)," *MIT Technology Review*, October 12, 2010, http://www.technologyreview.com/view/421187/why-japanese-love-robots-and-americans-fear-them/.

22 *There are more than 100 automation departments*: "List of Colleges That Offer Degree in Robotic Engineering," *Automation Components*, May 3, 2012, http://agi-automation.blogspot.com/2012/05/list-of-colleges-that-offer-degree-in.html; "List of Universities with a Robotics Program," National Aeronautics and Space Administration, http://robotics.nasa.gov/students/robo_u.php.

22 *In South Korea, teaching robots are seen*: "Comparison of Cultural Acceptability for Educational Robots between Europe and Korea," *Journal of Information Processing Systems* 4 (2008): 97–102, doi:10.3745/JIPS.2008.4.3.97.

22 *A recent study in the Middle East showed*: Nikolaos Mavridis, Marina-Selini Katsaiti, Silvia Naef, et al., "Opinions and Attitudes toward Humanoid Robots in the Middle East," *Springer Journal of AI and Society* 27 (2011): 517–34, http://www.academia.edu/1205802/Opinions_and_attitudes_toward_humanoid_robots_in_the_Middle_East.

23 *It is basically the application of algorithms*: Nick Cercone and Gordon McCalla, *The Knowledge Frontier: Essays in the Representation of Knowledge* (New York: Springer, 1987), 305.

23 *The recent exponential growth of robot data*: Ken Goldberg, "Cloud Robotics and Automation," UC Berkeley Current Projects, http://goldberg.berkeley.edu/cloud-robotics/.

24 *An imitation caterpillar robot designed by researchers*: John Schwartz, "In the Lab: Robots That Slink and Squirm," *New York Times*, March, 27, 2007, http://www.nytimes.com/2007/03/27/science/27robo.html?pagewanted=1&_r=1&ei=5070&en=91395fe7439a5b72&ex=1177128000.

24 *Apparently the Germans*: Alex Knapp, "The World's Largest Walking Robot Is a Giant Dragon," *Forbes*, September 18, 2013, http://www.forbes.com/sites/alexknapp/2013/09/18/the-worlds-largest-walking-robot-is-a-giant-dragon/.

25 *Run by the National Science Foundation*: "National Robotics Initiative Invests $38 Million in Next-Generation Robotics," *R&D Magazine*, October 25, 2013, http://www.rdmag.com/news/2013/10/national-robotics-initiative-invests-38-million-next-generation-robotics.

25 *The private sector is also investing*: John Markoff, "Google Adds to Its Menagerie of Robots," *New York Times*, December 14, 2013, http://www.nytimes.com/2013/12/14/technology/google-adds-to-its-menagerie-of-robots.html?_r=1&.

25　*As a kid, Hassabis was*: Samuel Gibbs, "Demis Hassabis: 15 Facts about the DeepMind Technologies Founder," *Guardian*, January 28, 2014, http://www.theguardian.com/technology/shortcuts/2014/jan/28/demis-hassabis-15-facts-deepmind-technologies-founder-google; "Breakthrough of the Year: The Runners-Up," *Science* 318, no. 5858 (2007): 1844–49, doi:10.1126/science.318.5858.1844a.

25　*At DeepMind, Demis and his colleagues*: "The Last AI Breakthrough Deep-Mind Made before Google Bought It for $400m," *Physics arXiv* (Blog), https://medium.com/the-physics-arxiv-blog/the-last-ai-breakthrough-deepmind-made-before-google-bought-it-for-400m-7952031ee5e1.

26　*It more than doubled in just three years*: Jennifer Hicks, "A New Series: The Future of Robotics, The Next 20 Years," *Forbes*, September 2, 2012, http://www.forbes.com/sites/jenniferhicks/2012/09/02/a-new-series-the-future-of-robotics-the-next-20-years/; Travis Deyle, "Venture Capital (VC) Funding for Robotics in 2014," *Hizook*, January 20, 2015, http://www.hizook.com/blog/2015/01/20/venture-capital-vc-funding-robotics-2014.

26　*In its first year of investment*: "The Next Big Thing," Grishin Robotics, http://grishinrobotics.com/#the_next_big_thing; Yuliya Chernova, "Robotics Investor Dmitry Grishin: The Future Is Happening," *Wall Street Journal*, July 23, 2013, http://blogs.wsj.com/venturecapital/2013/07/23/robotics-investor-dmitry-grishin-the-future-is-happening/.

26　*Singulariteam, a new Israeli venture capital fund*: Ingrid Lunden, "Israel VC Singulariteam Raises 2nd Fund, $102M Backed by Tencent, Renren Founders," *TechCrunch*, January 28, 2015, http://techcrunch.com/2015/01/28/singulariteam-vc-fund/n; Ingrid Lunden, "Meet Genesis Angels: A New $100M Fund for AI and Robotics, Co-Founded by Investor Kenges Rakishev and Chaired by Israel's Ex-PM," *TechCrunch*, April 19, 2013, http://techcrunch.com/2013/04/19/meet-genesis-angels-a-new-100m-fund-for-ai-and-robotics-from-investor-kenges-rakishev-and-led-by-israels-ex-pm/.

26　*The appeal for investors is obvious*: redazione, "Growth Forecast for Robotics Market to 2020," *Metalworking World Magazine*, June 3, 2014, http://www.metalworkingworldmagazine.com/growth-forecast-for-robotics-market-to-2020/.

26　*First, they argue that Moore's law*: Huw Price and Jaan Tallinn, "Artificial Intelligence: Can We Keep It in the Box?" *Conversation*, August 4, 2012, http://theconversation.com/artificial-intelligence-can-we-keep-it-in-the-box-8541.

27　*Those who argue for the singularity*: Lev Grossman, "2045: The Year Man Becomes Immortal," *Time*, February 10, 2011, http://content.time.com/time/magazine/article/0,9171,2048299,00.html; Paul Allen and Mark Greaves, "The Singularity Isn't Near," *MIT Technology Review*, October 12, 2011, http://www.technologyreview.com/view/425733/paul-allen-the-singularity-isnt-near.

27　*Those who argue against the possibility*: Ibid.

27　*Watson didn't actually "think"*: William Herkewitz, "Why Watson and Siri Are Not Real AI," *Popular Mechanics*, February 10, 2014, http://www

.popularmechanics.com/science/a3278/why-watson-and-siri-are-not-real-ai-16477207/.

27   *"Robots are going to become"*: Ken Goldberg, phone interview with Ari Ratner, October 4, 2013.

27   *Just as it would have been difficult*: "Statistics," YouTube, https://www.youtube.com/yt/press/statistics.html; "Follow the Audience . . . ," YouTube Official Blog, May 1, 2013, http://youtube-global.blogspot.com/2013/05/yt-brandcast-2013.html.

28   *General Motors introduced*: "The Original Futurama," *Wired*, November 27, 2007, http://www.wired.com/entertainment/hollywood/magazine/15-12/ff_futurama_original.

28   *Radar was a device on a hilltop*: Burkhard Bilger, "Auto-Correct," *New Yorker*, November 25, 2013, http://www.newyorker.com/reporting/2013/11/25/131125fa_fact_bilger?currentPage=2.

29   *As Sebastian Thrun explained*: "The Business and Culture of Our Digital Lives," *Los Angeles Times*, April 5, 2011, http://latimesblogs.latimes.com/technology/2011/04/googles-driverless-car-project-a-personal-one-for-engineer-sebastian-thrun.html.

29   *Worldwide, those statistics are enormous*: "Annual Global Road Crash Statistics," Association for Safe International Road Travel, http://asirt.org/Initiatives/Informing-Road-Users/Road-Safety-Facts/Road-Crash-Statistics.

30   *Accidents are caused by the four Ds*: Bilger, "Auto-Correct."

30   *There remain many gaps*: Lee Gomes, "Hidden Obstacles for Google's Self-Driving Cars," *MIT Technology Review*, August 28, 2014, http://www.technologyreview.com/news/530276/hidden-obstacles-for-googles-self-driving-cars/.

31   *Uber has already built*: John Biggs, "Uber Opening Robotics Research Facility in Pittsburgh to Build Self-Driving Cars," *TechCrunch*, February 2, 2015, http://techcrunch.com/2015/02/02/uber-opening-robotics-research-facility-in-pittsburgh-to-build-self-driving-cars/.

31   *At last count there were 162,037 active drivers*: Emily Badger, "Now We Know How Many Drivers Uber Has—and Have a Better Idea of What They're Making," *Washington Post*, January 22, 2015, http://www.washingtonpost.com/blogs/wonkblog/wp/2015/01/22/now-we-know-many-drivers-uber-has-and-how-much-money-theyre-making%E2%80%8B/.

31   *UPS and Google are also testing*: Salvador Rodriguez, "Amazon Is Not Alone: UPS, Google Also Testing Delivery Drones," *Los Angeles Times*, December 3, 2013, http://www.latimes.com/business/technology/la-fi-tn-amazon-ups-google-delivery-drones-20131203,0,3320223.story.

32   *In 2013, 1,300 surgical robots*: "World Robotics 2014 Service Robots," Service Robot Statistics, International Federation of Robotics, http://www.ifr.org/service-robots/statistics/.

32   *The number of robotic procedures*: Beth Howard, "Is Robotic Surgery Right for

You?" *AARP: The Magazine*, December 2013/January 2014, http://www.aarp.org/health/conditions-treatments/info-12-2013/robotic-surgery-risks-benefits.html.

32   *It's a minimally invasive remote robotic system*: "The Kindness of Strangers," *Economist*, January 8, 2012, http://www.economist.com/blogs/babbage/2012/01/surgical-robots.

33   *SEDASYS, already approved*: Jonathan Rockoff, "Robots vs. Anesthesiologists," *Wall Street Journal*, October 9, 2013, http://online.wsj.com/news/articles/SB10001424052702303983904579093252573814132.

33   *The* Journal for Healthcare Quality *has reported*: Roni Caryn Rabin, "New Concerns on Robotic Surgeries," *New York Times*, September 9, 2013, http://well.blogs.nytimes.com/2013/09/09/new-concerns-on-robotic-surgeries/?_r=0; Michol A. Cooper, Andrew Ibrahim, Heather Lyu, and Martin A. Makary, "Underreporting of Robotic Surgery Complications," *Journal for Healthcare Quality*, August 27, 2013, http://onlinelibrary.wiley.com/doi/10.1111/jhq.12036/abstract.

34   *He can raise his hand*: "Robots Allow Sick Children to Attend School 'in Person,'" *KHOU.com*, May 10, 2013, http://www.khou.com/story/news/local/2014/07/23/12045114/.

34   *a less than two-foot-tall*: "Who Is NAO?" Aldebaran Robotics, http://www.aldebaran.com/en/humanoid-robot/nao-robot.

34   *It has also been adapted*: "Robots Being Used as Classroom Buddies for Children with Autism," University of Birmingham, November 8, 2012, http://www.birmingham.ac.uk/news/latest/2012/11/8-Nov-Robots-being-used-as-classroom-buddies-for-children-with-autism.aspx.

35   *At an elementary school in Harlem*: "Teaching, With Help From a Robot," *Wall Street Journal*, video, April 10, 2013, http://www.wsj.com/video/teaching-with-help-from-a-robot/B5775430-2A00-4397-9EC9-A3B0877FF908.html#!B5775430-2A00-4397-9EC9-A3B0877FF908; Sandra Okita, bio page, Teachers College, Columbia University, http://www.tc.columbia.edu/academics/?facid=so2269.

35   *Jellyfish cost the world's fishing*: Lynne Peeples, "Jellyfish Stings an Increasing Public Health Concern, Experts Say," *Huffington Post*, October 19, 2013, http://www.huffingtonpost.com/2013/10/19/jellyfish-stings-increasing-health_n_4122006.html; "These Robots Hunt Jellyfish—and Then Liquify Them with Rotating Blades of Death," *Co.Exist*, October 3, 2013, http://www.fastcoexist.com/3019164/these-robots-hunt-jellyfish-and-then-liquify-them-with-rotating-blades-of-death.

35   *Then the Urban Robotics Lab*: Drew Prindle, "Meet South Korea's Autonomous Jellyfish-Murdering Robots," *Digital Trends*, October 8, 2013, http://www.digitaltrends.com/cool-tech/jellyfish-murdering-robots/.

36   *Perhaps thinking ahead about both the economics*: Lee Chyen Yee and Clare Jim, "Foxconn to Rely More on Robots; Could Use 1 Million in 3 Years," Reuters, August 1, 2011, http://www.reuters.com/article/2011/08/01/us-foxconn-robots-idUSTRE77016B20110801; Tiffany Kaiser, "Foxconn Receives

10,000 Robots to Replace Human Factory Workers," *Daily Tech*, November 4, 2012, http://www.dailytech.com/Foxconn+Receives+10000+Robots+to+R eplace+Human+Factory+Workers+/article29194.htm; Philip Elmer-DeWitt, "By the Numbers: How Foxconn Churns Out Apple's iPhone 5S," *Fortune*, November 27, 2013, http://tech.fortune.cnn.com/2013/11/27/apple-foxconn-factory-iphone/.

36 *Each of these robots currently costs $25,000*: John Biggs, "Foxconn Allegedly Replacing Human Workers with Robots," *TechCrunch*, November 13, 2012, http://techcrunch.com/2012/11/13/foxconn-allegedly-replacing-human-workers-with-robots/; Nicholas Jackson, "Foxconn Will Replace Workers with 1 Million Robots in 3 Years," *Atlantic*, July 31, 2011, http://www.theatlantic.com/technology/archive/2011/07/foxconn-will-replace-workers-with-1-million-robots-in-3-years/242810/.

36 *By the end of 2012*: Jackson, "Foxconn Will Replace Workers."

36 *Gou hopes to have the first*: Robert Skidelsky, "Rise of the Robots: What Will the Future of Work Look Like?" *Guardian*, February 19, 2013, http://www.theguardian.com/business/2013/feb/19/rise-of-robots-future-of-work.

37 *As he explained in a 2012* New York Times *article*: John Markoff, "Skilled Work, without the Worker," *New York Times*, August 19, 2012, http://www.nytimes.com/2012/08/19/business/new-wave-of-adept-robots-is-changing-global-industry.html?pagewanted=all&_r=0.

37 *But wages in China*: Keith Bradsher, "Even as Wages Rise, China Exports Grow," *New York Times*, January 10, 2014, http://www.nytimes.com/2014/01/10/business/international/chinese-exports-withstand-rising-labor-costs.html?hpw&rref=business.

38 *During the recent recession*: Erik Brynjolfsson and Andrew McAfee, *Race against the Machine: How the Digital Revolution Is Accelerating Innovation, Driving Productivity, and Irreversibly Transforming Employment and the Economy* (Lexington, MA: Digital Frontier, 2011).

38 *Two Oxford University professors*: Carl Benedikt Frey and Michael A. Osborne, "The Future of Employment: How Susceptible Are Jobs to Computerisation?" Oxford Martin School, 2013, http://www.oxfordmartin.ox.ac.uk/downloads/academic/The_Future_of_Employment.pdf.

39 *It measures the shape and size of the customer's head*: Ibid.

39 *By way of illustration*: "Reinventing Low Wage Work: The Restaurant Workforce in the United States," Aspen Institute, October 30, 2014, http://www.aspenwsi.org/wordpress/wp-content/uploads/The-Restaurant-Workforce-in-the-United-States.pdf.

39 *More than 2.3 million people are*: "Occupational Employment Statistics: Occupational Employment and Wages, May 2014," Bureau of Labor Statistics, March 25, 2015, http://www.bls.gov/oes/current/oes353031.htm.

40 *These robots, designed by the Japanese company Motoman*: "About Us: About Robots," Hajime Robot Restaurant, http://hajimerobot.com.

40 *Currently youth unemployment*: Zachary Karabell, "The Youth

Unemployment Crisis Might Not Be a Crisis," *Atlantic*, November 25, 2013, http://www.theatlantic.com/business/archive/2013/11/the-youth-unemployment-crisis-might-not-be-a-crisis/281802/.

40 *People are falling behind*: David Rotman, "How Technology Is Destroying Jobs," *MIT Technology Review*, June 12, 2013, http://www.technologyreview.com/featuredstory/515926/how-technology-is-destroying-jobs/.

41 *In 1950, 13 percent of China's population*: Ian Johnson, "China's Great Uprooting: Moving 250 Million into Cities," *New York Times*, June 15, 2013, http://www.nytimes.com/2013/06/16/world/asia/chinas-great-uprooting-moving-250-million-into-cities.html.

41 *By comparison, the United States has*: "Metropolitan Areas: Assessing Competitive Position and Change," ProximityOne, http://proximityone.com/metros2013.htm.

CHAPTER 2: THE FUTURE OF THE HUMAN MACHINE

45 *Survival rates for a first relapse are slim*: "Doctor Survives Cancer He Studies," McDonnell Genome Institute, Washington University, http://genome.wustl.edu/articles/detail/doctor-survives-cancer-he-studies.

45 *They decided to do something*: Lukas Wartman, Skype interview with Teal Pennebaker, December 2, 2013.

46 *It turned out that one*: Gina Kolata, "In Treatment for Leukemia, Glimpses of the Future," *New York Times*, July 7, 2012, http://www.nytimes.com/2012/07/08/health/in-gene-sequencing-treatment-for-leukemia-glimpses-of-the-future.html?pagewanted=1.

46 *Four years later*: Wartman, interview.

47 *But the breakthrough that launched*: "*Haemophilus influenzae* Disease (Including Hib)," Centers for Disease Control and Prevention, http://www.cdc.gov/hi-disease/.

47 *If we could unravel*: "The Human Genome Project Completion: Frequently Asked Questions," National Human Genome Institute, October 30, 2010, http://www.genome.gov/11006943.

48 *The cost of mapping*: Ibid.

48 *Lander helped sequence the human genome*: "Eric S. Lander," Broad Institute, https://www.broadinstitute.org/history-leadership/scientific-leadership/core-members/eric-s-lander.

48 *The size of the genomics market*: "Genomics Market by Products—[Instruments (NGS platform, Microarray, RT-PCR), Consumables (Genechips, Reagents for DNA Extraction & Purification, Sequencing)], Services (Sequencing & Microarray Services, and Software)—Global Forecast to 2018," *Markets and Markets*, January 2014, http://www.marketsandmarkets.com/Market-Reports/genomics-market-613.html.

48 *Ronald W. Davis, director*: Forbes Leadership Forum: Ronald W. Davis, "It's Time to Bet on Genomics," *Forbes*, June 1, 2012, http://www.forbes.com/sites/forbesleadershipforum/2012/06/01/its-time-to-bet-on-genomics/.

48   *He's also one of the most*: "Bert Vogelstein," Nobelprize.org, Nobel Media AB 2014, http://www.nobelweekdialogue.org/participants/vogelstein/.

49   *In the past 40 years*: "Investigative Instincts Guided Vogelstein's Journey of Discovery," OncLive, September 12, 2014, http://www.onclive.com/publications/ Oncology-live/2014/August-2014/Investigative-Instincts-Guided-Vogelsteins-Journey-of-Discovery; "Essential Science Indicators," Thomson Reuters, 2014, http://thomsonreuters.com/essential-science-indicators/.

49   *In the 1980s, Vogelstein*: Eric R. Fearon, Stanley R. Hamilton, and Bert Vogelstein, "Clonal Analysis of Human Colorectal Tumors," *Science* 238, no. 4824 (1987): 193–97, http://www.ncbi.nlm.nih.gov/pubmed/2889267.

49   *The amount can be so small*: Antonio Regalado, "Spotting Cancer in a Vial of Blood," *MIT Technology Review*, August 11, 2014, http://www.technologyreview.com/featuredstory/529911/spotting-cancer-in-a-vial-of-blood/.

49   *Vogelstein says*: Ibid.

50   *But by stage 4*: "Types and Stages of Ovarian Cancer," National Ovarian Cancer Coalition, http://www.ovarian.org/types_and_stages.php.

50   *Better genetic diagnostic testing*: Luis Diaz, interview with Teal Pennebaker, November 19, 2013.

50   *So in 2009, Diaz*: "Advisors," Personal Genome Diagnostics, http://main.personalgenome.com/advisors/.

51   *Once in the machine*: PGDx team and tour of facilities by Teal Pennebaker, Baltimore, MD, December 2013.

51   *But often the right*: Ibid.

52   *This revolutionary possibility got*: "Fact Sheet: President Obama's Precision Medicine Initiative," White House, January 30, 2015, https://www.whitehouse.gov/the-press-office/2015/01/30/fact-sheet-president-obama-s-precision-medicine-initiative; Jocelyn Kaiser, "NIH Plots Million-Person Megastudy," *Science* 347, no. 6224 (2015): 817, http://www.sciencemag.org/content/347/6224/817.summary?utm_source=twitter&utm_medium=social&utm_campaign=twitter.

53   *"Seeking help is a sign"*: Josh Rogin, "Clinton to State Employees: Seek Mental Health Help If You Need It," *Foreign Policy*, September 10, 2010, http://thecable.foreignpolicy.com/posts/2010/09/10/clinton_to_state_employees_seek_mental_health_help_if_you_need_it.

53   *Psychotherapy was the most common*: Richard G. Frank and Sherry Glied, *Better But Not Well: Mental Health Policy in the United States since 1950* (Baltimore: Johns Hopkins University Press, 2006), 764.

54   *Side effects ran the gamut*: J. A. Lieberman, "History of the Use of Antidepressants in Primary Care," *Journal of Clinical Psychiatry* 5, no. 7 (2003): 6–9, http://www.psychiatrist.com/pcc/pccpdf/v05s07/v05s0702.pdf.

54   *Prozac, the first of these*: Anna Moore, "Eternal Sunshine," *Guardian*, May 13, 2007, http://www.theguardian.com/society/2007/may/13/socialcare.medicineandhealth.

54   *Fifteen years after it hit*: Laura Fitzpatrick, "A Brief History of

Antidepressants," *Time*, January 7, 2010, http://content.time.com/time/health/article/0,8599,1952143,00.html.

54 *By 2008, antidepressants*: Siddhartha Mukherjee, "Post-Prozac Nation," *New York Times*, April 22, 2012, http://www.nytimes.com/2012/04/22/magazine/the-science-and-history-of-treating-depression.html?ref=prozacdrug; Qiuping Gu, Charles F. Dillon, and Vicki L. Burt, "Prescription Drug Use Continues to Increase: US Prescription Drug Data for 2007–2008," NCHS data brief, September 2010, http://www.cdc.gov/nchs/data/databriefs/db42.pdf.

54 *Today most medical treatments*: Ray DePaulo, interview with Teal Pennebaker, December 9, 2013.

55 *One interesting opportunity is*: Johns Hopkins Medical Institutions, "Genetic Link to Attempted Suicide Identified," *ScienceDaily*, http://www.sciencedaily.com/releases/2011/03/110328131258.htm.

55 *Uncle Ray's colleagues at Johns Hopkins*: V. L. Willour, F. Seifuddin, P. B. Mahon, et al., "A Genome-Wide Association Study of Attempted Suicide," *Molecular Psychiatry* 17 (2012): 433–44, http://www.ncbi.nlm.nih.gov/pubmed/21423239.

56 *Fetal DNA tests have*: "Down Syndrome: Tests and Diagnosis," Mayo Clinic, http://www.mayoclinic.org/diseases-conditions/down-syndrome/basics/tests-diagnosis/con-20020948.

57 *Bert Vogelstein and Luis Diaz are concerned*: Bert Vogelstein, interview with Teal Pennebaker, December 9, 2013.

57 *Founded by Anne Wojcicki*: Katie Hafner, "Silicon Valley Wide-Eyed over a Bride," *New York Times*, May 29, 2007, http://www.nytimes.com/2007/05/29/technology/29google.html.

57 *the company provides ancestry-related*: "How It Works," *23andMe*, https://www.23andme.com/howitworks/.

57 *It's not a full sequencing*: "About the 23andMe Personal Genome Service," 23andMe, https://customercare.23andme.com/entries/22591668.

58 *Since then, he drinks green tea*: Elizabeth Murphy, "Do You Want to Know What Will Kill You?" *Salon*, October 25, 2013, http://www.salon.com/2013/10/25/inside_23andme_founder_anne_wojcickis_99_dna_revolution_newscred/.

58 *all of them have faced*: Kira Peikoff, "I Had My DNA Picture Taken, with Varying Results," *New York Times*, December 30, 2013, http://www.nytimes.com/2013/12/31/science/i-had-my-dna-picture-taken-with-varying-results.html?src=recg.

58 *In late 2013, it demanded*: Chris O'Brien, "23andMe Suspends Health-Related Genetic Tests after FDA Warning," *Los Angeles Times*, December 6, 2013, http://articles.latimes.com/2013/dec/06/business/la-fi-tn-23andme-suspends-tests-fda-20131205.

58 *The FDA's public letter*: "23andMe, Inc. 11/22/13," FDA: Inspections, Compliance, Enforcement, and Criminal Investigation Warning Letters, November 22, 2013, http://www.fda.gov/iceci/enforcementactions/

warningletters/2013/ucm376296.htm; Scott Hensley, "23andMe Bows to FDA's Demands, Drops Health Claims," National Public Radio, December 6, 2013, http://www.npr.org/blogs/health/2013/12/06/249231236/23andme-bows-to-fdas-demands-drops-health-claims.

58 *Now their tests promise only*: Ibid.

59 *At this time we do not*: "How It Works."

59 *Through a partnership*: "Michael J. Fox, Our Big-Time Hero," 23andMe, April 27, 2012, http://blog.23andme.com/news/inside-23andme/michael-j-fox-our-big-time-hero/; Matthew Herper, "Surprise! With $60 Million Genentech Deal, 23andMe Has a Business Plan," *Forbes*, January 6, 2015, http://www.forbes.com/sites/matthewherper/2015/01/06/surprise-with-60-million-genentech-deal-23andme-has-a-business-plan/.

60 *Its signature product, Genophen, sequences*: "Our Model," Genophen: How It Works, http://www.genophen.com/consumers/how-it-works/our-model; Davis, "It's Time to Bet on Genomics."

61 *The doctors access the Genophen*: "This Startup Will Make You a Personalized Health Plan Based on Your Genes," *Co.Exist*, July 8, 2014, http://www.fastcoexist.com/3032567/this-startup-will-make-you-a-personalized-health-plan-based-on-your-genes.

61 *Three years ago*: PGDx team and tour of facilities.

62 *Venter, the second of four*: Ross Douthat, "The God of Small Things," *Atlantic*, January/February 2007, http://www.theatlantic.com/magazine/archive/2007/01/the-god-of-small-things/305556/.

62 *He was both motivated*: Meredith Wadman, "Biology's Bad Boy Is Back. Craig Venter Brought Us the Human Genome. Now He Aims to Build a Life Form That Will Change the World." *Fortune*, March 8, 2004, http://archive.fortune.com/magazines/fortune/fortune_archive/2004/03/08/363705/index.htm.

62 *When he asked the head*: Victor K. McElheny, *Drawing the Map of Life: Inside the Human Genome Project* (New York: Basic Books, 2010), 96.

62 *The initial focus is on lungs*: Bradley J. Fikes, "Modified Pigs to Grow Humanized Lungs," *San Diego Union-Tribune*, May 6, 2014, http://www.utsandiego.com/news/2014/may/06/synthetic-genomics-pigs-lung-therapeutics/.

63 *Today, with the emergence*: Human Longevity Inc., http://www.humanlongevity.com/.

63 *HLI has recently lined up*: Sarah Gantz, "Human Genome Pioneer J. Craig Venter Taps Baltimore Startup for Next Project," *Baltimore Business Journal*, January 12, 2015, http://www.bizjournals.com/baltimore/blog/cyberbiz-blog/2015/01/human-genome-pioneer-j-craig-ventertaps-baltimore.html.

63 *It raised $70 million in venture capital*: Bryan Johnson, Twitter post, October 21, 2014, https://twitter.com/bryan_johnson/status/524628698842951680.

63 *A few years back*: Carl Zimmer, "Bringing Them Back to Life," *National Geographic*, April 2013, http://ngm.nationalgeographic.com/2013/04/125-species-revival/zimmer-text.

63 *Granted, the bucardo didn't live long*: Nathaniel Rich, "The Mammoth

Cometh," *New York Times Magazine*, March 2, 2014, http://www.nytimes .com/2014/03/02/magazine/the-mammoth-cometh.html?ref=magazine&_ r=0.

64 *In 2012, the Revive & Restore project*: "Revive and Restore," https://www .facebook.com/ReviveandRestoreProject/info.

64 *As Revive & Restore sees it*: "Revive and Restore," Long Now Foundation, http://longnow.org/revive/.

64 *As with the bucardo*: Rich, "The Mammoth Cometh."

64 *Efforts are already under way*: Ibid.

65 *The United States is the number one*: Richard van Noorden, "Global Mobility: Science on the Move," *Nature*, October 17, 2012, http://www.nature .com/news/global-mobility-science-on-the-move-1.11602.

65 *No longer just a 1 percent*: Eric J. Topol, "Gore on the Genomics Race with China: Is the US Losing?" *Medscape*, March 7, 2014, http://www.medscape .com/viewarticle/821001.

65 *Some of its researchers are*: Al Gore, *The Future: Six Drivers of Global Change* (New York: Random House, 2013).

65 *Since 1998, the share of China's economy*: Richard van Noorden, "China Tops Europe in R&D Intensity," *Nature*, January 8, 2014, http://www.nature .com/news/china-tops-europe-in-rd-intensity-1.14476.

65 *While the portion*: David Wertime, "It's Official: China Is Becoming a New Innovation Powerhouse," *Foreign Policy*, February 7, 2014, http://www .foreignpolicy.com/articles/2014/02/06/its_official_china_is_becoming_a_ new_innovation_powerhouse.

66 *While China's investment is climbing*: National Science Foundation, "Research and Development: National Trends and International Comparisons," in *Science and Engineering Indicators 2014*, http://www.nsf.gov/statistics/ seind14/index.cfm/chapter-4/c4h.htm.

66 *Meanwhile, China's output has been*: National Science Foundation, "Academic Research and Development," in *Science and Engineering Indicators 2014*, http://www.nsf.gov/statistics/seind14/index.cfm/chapter-5.

66 *China's State Council has established*: Gore, *The Future*.

66 *In three years, the Chinese*: Topol, "Gore on the Genomics Race with China."

67 *Other sources include providing data*: Michael Specter, "The Gene Factory," *New Yorker*, January 6, 2014, http://archives.newyorker.com/?i=2014- 01-06#folio=036; Christina Larson, "Inside China's Genome Factory," *MIT Technology Review*, February 11, 2013, http://www.technologyreview.com/ featuredstory/511051/inside-chinas-genome-factory/.

68 *While the Soviet system produced huge numbers*: Anthony Ramirez, "World-Class Research, for a Song," *New York Times*, January 11, 1993, http://www .nytimes.com/1993/01/11/business/world-class-research-for-a-song.html.

68 *In Lysenko's scientific view*: Peter Ferrara, "The Disgraceful Episode of Lysenkoism Brings Us Global Warming Theory," *Forbes*, April 28, 2013, http://www.forbes.com/sites/peterferrara/2013/04/28/

the-disgraceful-episode-of-lysenkoism-brings-us-global-warming-theory/; Rodney Shackleford, "Trofim Lysenko, Soviet Ideology, and Pseudo-Science," *h+ Magazine*, May 22, 2013, http://hplusmagazine.com/2013/05/22/trofim-lysenko-soviet-ideology-and-pseudo-science/.

68 *Lysenko convinced the Soviet Agriculture Academy*: Jacob Darwin Hamblin, *Science in the Early Twentieth Century: An Encyclopedia* (Santa Barbara, CA: ABC-CLIO, 2005), 188–89.

68 *Those who embraced Lysenko-style research*: Ferrara, "The Disgraceful Episode of Lysenkoism Brings Us Global Warming Theory."

68 *On the flip side, scientists*: Shackleford, "Trofim Lysenko."

69 *The first "ethnically Russian" genome*: Kevin Davies, "The Russians Are Coming: Moscow Institute Sequences First 'Ethnically Russian' Genome," *Bio-ITWorld*, May 14, 2010, http://www.bio-itworld.com/news/05/14/10/Russian-institute-sequences-ethnically-russian-genome.html.

69 *Six billion of the 7 billion*: "Deputy UN Chief Calls for Urgent Action to Tackle Global Sanitation Crisis," *UN News Centre*, March 21, 2013, http://www.un.org/apps/news/story.asp?NewsID=44452#.VFOp5PTF800.

69 *During my travels through Africa and low-income*: Karin Källander, James K. Tibenderana, Onome J. Akpogheneta, et al., "Mobile Health (mHealth) Approaches and Lessons for Increased Performance and Retention of Community Health Workers in Low- and Middle-Income Countries: A Review," *Journal of Medical Internet Research*, January 25, 2013, http://www.ncbi.nlm.nih.gov/pmc/articles/PMC3636306/.

70 *Community health workers frequently walked*: "Josh Nesbit," *Forbes* Medic Profile, http://www.forbes.com/impact-30/josh-nesbit.html; "Our Story," Medic Mobile, http://medicmobile.org/team.

70 *The World Health Organization estimates*: Nadim Mahmud, Joce Rodriguez, and Josh Nesbit, "A Text Message–Based Intervention to Bridge the Healthcare Communication Gap in the Rural Developing World," *Technology and Healthcare* 18 (2010): 137–44, http://www.researchgate.net/publication/44623382_A_text_message–based_intervention_to_bridge_the_healthcare_communication_gap_in_the_rural_developing_world.

71 *A Kenyan company, Shimba*: "Kenya in Numbers," *mwakilishi*, June 1, 2012, http://www.mwakilishi.com/content/articles/2012/06/01/kenya-in-numbers.html.

71 *The World Bank puts the doctor*: World Bank, "Physicians (per 1,000 People)," in *World Development Indicators*, http://data.worldbank.org/indicator/SH.MED.PHYS.ZS.

71 *To address the doctor deficit*: Gabriel Demombynes and Aaron Thegeya, "Kenya's Mobile Revolution and the Promise of Mobile Savings," Policy Research working papers, March 2012, http://elibrary.worldbank.org/doi/book/10.1596/1813-9450-5988.

71 *The app has a symptom checker*: Nicolas Friederici, Carol Hullin, and Masatake Yamamichi, "mHealth," World Bank 2012 Information and

Communications for Development Report, http://siteresources.worldbank
.org/EXTINFORMATIONANDCOMMUNICATIONANDTECHNOLOGIES/
Resources/IC4D-2012-Chapter-3.pdf.

71 *In a country with vast rural areas*: Franco Papeschi, "Problem: 7,000 Doctors Serve a Nation of 40 Million People. Solution: MedAfrica," World Wide
Web Foundation, March 12, 2012, http://www.webfoundation.org/2012/03/
medafrica-interview/.

72 *This saves a trip to*: Bill Bulkeley, "Your Phone Can Take Your Blood Pressure with This New Tech," *Forbes*, November 11, 2013, http://www.forbes
.com/sites/ptc/2013/11/11/your-phone-can-take-your-blood-pressure-with-
this-new-tech/.

72 *Shortly after launching, it successfully*: CrunchBase: EyeNetra, http://www
.crunchbase.com/organization/eyenetra.

72 *They could look at dozens*: "Mammograms," National Cancer Institute,
http://www.cancer.gov/cancertopics/factsheet/detection/mammogram.

73 *Most health insurance companies pay*: Jeanne Pinder, "How Much Does a
Mammogram Cost? Our Survey with WNYC: $0 to $2,786.95!" Clear Health
Costs, May 22, 2013, http://clearhealthcosts.com/blog/2013/05/how-much-
does-a-mammogram-cost-prices-payments-vary-widely-our-survey-with-
wnyc-finds/.

CHAPTER 3: THE CODE-IFICATION OF MONEY, MARKETS, AND TRUST

76 *The "peso," the Israeli "shekel"*: "peso," Online Etymology Dictionary,
http://www.etymonline.com/index.php?term=peso&allowed_in_frame=0;
"shekel," Online Etymology Dictionary, http://www.etymonline.com/index
.php?term=shekel&allowed_in_frame=0; "pound," Online Etymology Dictionary, http://www.etymonline.com/index.php?term=pound&allowed_in_frame=0.

76 *"Ruble" comes from the Old Russian*: "ruble," Online Etymology Dictionary,
http://www.etymonline.com/index.php?term=ruble&allowed_in_frame=0.

77 *All that was required was*: Mary Bellis, "Automatic Teller Machines—ATM,"
About.com, http://inventors.about.com/od/astartinventions/a/atm.htm.

77 *The online payments service*: PayPal History, "PayPal: About Us," https://
www.paypal-media.com/au/history.

77 *More than half of American adults*: Susannah Fox, "51% of US Adults Bank
Online," Pew Research Center: Internet, Science and Tech, August 7, 2013,
http://www.pewinternet.org/2013/08/07/51-of-u-s-adults-bank-online/;
"Mobile Banking Users to Exceed 1.75 Billion by 2019, Representing 32%
of the Global Adult Population," Juniper Research, July 8, 2014, http://www
.juniperresearch.com/viewpressrelease.php?pr=356.

77 *By 2017, that number*: "Study: Mobile Banking Users to Exceed 1 Billion
Worldwide by 2017," ATMmarketplace.com, January 9, 2013, http://www
.atmmarketplace.com/article/206411/Study-Mobile-banking-users-to-
exceed-1-billion-worldwide-by-2017.

79 *The company made world-class glass*: D. T. Max, "Two-Hit Wonder," *New Yorker*,

October 21, 2013, http://www.newyorker.com/reporting/2013/10/21/131021fa_fact_max?currentPage=all.

79 *They released their first product*: Jeremy Horwitz, "Review: Square, Inc. Square Credit Card Reader (Second-Generation)," *iLounge*, March 28, 2011, http://www.ilounge.com/index.php/reviews/entry/square-inc.-square-credit-card-reader-second-generation/; Rachel King, "Jack Dorsey: Square Has Processed 1 Billion Payments," *ZDNet*, November 6, 2014, http://www.zdnet.com/jack-dorsey-square-has-processed-1-billion-payments-7000035529/.

79 *Square is part of a fierce competition*: Matt Weinberger, "Here's the Next Key Challenge for Stripe, the Hot Payment Startup Whose Valuation Keeps Soaring," *BusinessInsider*, May 21, 2015, http://www.businessinsider.com/stripe-valuation-hitting-5-billion-as-payments-market-heats-up-2015-5.

79 *These fees add up*: Sara Angeles, "How to Accept Credit Cards Online, In-Store or Anywhere: 2015 Guide," *Business News Daily*, June 11, 2015, http://www.businessnewsdaily.com/4394-accepting-credit-cards.html.

80 *And credit card companies do not*: John Tozzi, "Merchants Seek Lower Credit Card Interchange Fees," *Bloomberg Businessweek*, October 7, 2009, http://www.businessweek.com/smallbiz/running_small_business/archives/2009/10/merchants_seek.html.

82 *Each year, 2,000 Palestinians graduate*: Alec Ross, "Light Up the West Bank," *Foreign Policy*, June 18, 2013, http://www.foreignpolicy.com/articles/2013/06/18/why_the_west_bank_needs_3g.

82 *As a result, the service is banned*: John Lister, "What Country Doesn't Work with PayPal?" *Houston Chronicle: Small Business*, http://smallbusiness.chron.com/country-doesnt-work-paypal-66099.html; "Feasibility Study: Microwork for the Palestinian Territories," World Bank: Country Management Unit for the Palestinian Territories (MNC04) and the Information and Communication Technologies Unit, February 2013, http://siteresources.worldbank.org/INTWESTBANKGAZA/Resources/Finalstudy.pdf.

82 *Among the first of these was Alibaba*: "The World's Greatest Bazaar," *Economist*, May 23, 2013, http://www.economist.com/news/briefing/21573980-alibaba-trailblazing-chinese-internet-giant-will-soon-go-public-worlds-greatest-bazaar.

83 *Its Alipay payment system*: Avi Mizrahi, "Alipay Set for IPO after Alibaba Brought in a Record $9.3 Billion in 24 Hours on Singles Day," *Finance Magnates*, November 12, 2014, http://forexmagnates.com/alipay-set-ipo-alibaba-brought-record-9-3-billion-24-hours-singles-day/.

83 *Driven by competition over natural resources*: Nicholas Kristof, "The Pain of the G-8's Big Shrug," *New York Times*, July 10, 2008, http://www.nytimes.com/2008/07/10/opinion/10kristof.html?_r=0.

83 *As security collapsed, most foreign*: "Democratic Republic of the Congo: Economy," Michigan State University globalEDGE, http://globaledge.msu.edu/countries/democratic-republic-of-the-congo/economy.

83 *At least 75 percent*: "Democratic Republic of Congo: Country Plan,"

Department for International Development, 2008, http://www.oecd.org/countries/democraticrepublicofthecongo/40692153.pdf.

83 *A third of the population*: "Human Development Indicators," United Nations Development Programme, http://hdr.undp.org/en/countries/profiles/GNQ.html.

84 *Children walked barefoot*: Sudarsan Raghavan, "In Traumatic Arc of a Refugee Camp, Congo's War Runs Deep," *Washington Post*, November 7, 2013, http://www.washingtonpost.com/world/africa/in-traumatic-arc-of-a-refugee-camp-congos-war-runs-deep/2013/11/07/22de1dbe-470b-11e3-95a9-3f15b5618ba8_story.html.

84 *The mobile penetration rate*: "Democratic Republic of Congo—Telecoms, Mobile and Broadband—Market Insights and Statistics," *Market Briefing*, October 2014, http://www.telecomsmarketresearch.com/research/TMAABOEF-Democratic-Republic-of-Congo---Telecoms--Mobile-and-Broadband---Market-Insights-and-Statistics.shtml.

85 *What goes for the Congo is*: Matt Twomey, "Cashless Africa: Kenya's Smash Success with Mobile Money," CNBC, November 11, 2013, http://www.cnbc.com/id/101180469.

85 *Today that number is over*: John Koetsier, "African Mobile Penetration Hits 80% (and Is Growing Faster Than Anywhere Else)," *VentureBeat*, December 3, 2013, http://venturebeat.com/2013/12/03/african-mobile-penetration-hits-80-and-is-growing-faster-than-anywhere-else/.

86 *By 2012, 19 million M-Pesa*: "Is It a Phone, Is It a Bank?" *Economist*, March 27, 2013, http://www.economist.com/news/finance-and-economics/21574520-safaricom-widens-its-banking-services-payments-savings-and-loans-it.

86 *While estimates vary, the adoption*: "On the New Frontier of Mobile and Money in the Developing World: Mobile Phones, M-PESA, and Kenya," *Hydra: Interdisciplinary Journal of Social Sciences* 1, no. 2 (2013): 49–59.

86 *Anyone with a valid identification*: Gabriel Demombynes and Aaron Thegeya, "Kenya's Mobile Revolution and the Promise of Mobile Savings," Policy Research working papers, March 2012, http://elibrary.worldbank.org/doi/abs/10.1596/1813-9450-5988.

86 *Withdrawing money is just as easy*: "What is Mpesa? How Does It Work? How Did It Start?" OurMobileWorld.org, January 1, 2012, http://ourmobileworld.org/post/35349373601/what-is-mpesa-how-does-it-work-how-did-it-start.

87 *The process is safe, since M-Pesa*: "Dial M for Money," *Economist*, June 28, 2007, http://www.economist.com/node/9414419.

87 *M-Shwari also facilitates the disbursement*: Ignacio Mas and Tonny Omwansa, "NexThought Monday: A Close Look at Safaricom's M-Shwari," next billion, December 10, 2012, http://www.nextbillion.net/blogpost.aspx?blogid=3050.

87 *In another program, M-Pesa works*: "Deacons Kenya Customers to Pay Via M-Pesa," Safaricom, March 15, 2011, http://www.safaricom.co.ke/personal/m-pesa/m-pesa-resource-centre?layout=edit&id=437.

87   *Approximately $40 billion is sent*: Sanket Mohaprata and Dilip Ratha, *Remittance Markets in Africa* (Washington, DC: The World Bank, 2011), http://siteresources .worldbank.org/EXTDECPROSPECTS/Resources/476882-1157133580628/ RMA_FullReport.pdf; "Remittances Gaining Increasing Importance in Africa: New Report from the African Development Bank," SilverStreet Capital, July 22, 2013, http://www.silverstreetcapital.com/Publisher/File.aspx?ID=112754.

87   *And there are additional problems*: "Remittance Fees Hurt Africans, Says Comic Relief," *BBC News*, April 16, 2014, http://www.bbc.com/news/business-27046285.

87   *They, like Square, Stripe, and Apple Pay*: Nye Longman, "$33 billion Says Africa Is Still the Mobile Continent," *African Business Review*, May 21, 2015, http://www.africanbusinessreview.co.za/technology/1947/33-billion-says-Africa-is-still-the-mobile-continent.

87   *Although most mobile payments systems*: "Chapter 4: Remittances," United Nations Development Programme: *Towards Human Resilience: Sustaining MDG Progress in an Age of Economic Uncertainty*, September 2011, http://www.undp.org/content/dam/undp/library/Poverty%20Reduction/ Inclusive%20development/Towards%20Human%20Resilience/Towards_ SustainingMDGProgress_Ch4.pdf.

87   *M-Pesa is not far*: Loek Essers, "MoneyGram and Vodafone M-Pesa Bring Mobile Remittances to New Countries," *PCWorld*, February 11, 2014, http:// www.pcworld.com/article/2096620/moneygram-and-vodafone-mpesa-bring-mobile-remittances-to-new-countries.html.

88   *In comparison, remittance fees*: "Sending Money from United States to Kenya," The World Bank: Remittance Prices Worldwide, April 27, 2015, http://remittanceprices.worldbank.org/en/corridor/United-States/ Kenya.

88   *Indeed, Mo had put his money*: "MTC Acquires Celtel International B.V.," *Zain*, March 29, 2005, http://www.zain.com/media-center/press-releases/ mtc-acquires-celtel-international-bv/.

89   *And in 2007, he created*: "Prize Offered to Africa's Leaders," *BBC News*, October 26, 2006, http://news.bbc.co.uk/2/hi/uk_news/6086088.stm.

89   *In the seven years since*: Mo Ibrahim, "Celtel's Founder on Building a Business on the World's Poorest Continent," *Harvard Business Review*, October 2012, http://hbr.org/2012/10/celtels-founder-on-building-a-business-on-the-worlds-poorest-continent/ar/1; "Mo Ibrahim African Leaders Prize Unclaimed Again," *BBC News*, October 14, 2013, http://www.bbc.com/news/world-africa-24521870.

90   *eBay's business is based*: "Online Extra: Pierre Omidyar on 'Connecting People,' *Bloomberg Businessweek*, June 19, 2005, http://www.businessweek .com/printer/articles/195874-online-extra-pierre-omidyar-on-connecting-people?type=old_article.

91   *It has more than 800,000 listings*: "About Us," Airbnb, https://www.airbnb .com/about/about-us.

91   *With a valuation of $20 billion*: Ingrid Lunden, "Airbnb Is Raising a Monster Round at a $20B Valuation," *TechCrunch*, February 27, 2015, http://techcrunch .com/2015/02/27/airbnb-2/; "Hyatt Hotels Corporation (H)," *Yahoo! Finance*, http://finance.yahoo.com/q?s=H; "#1006 Brian Chesky," "The World's Billionaires," *Forbes*, http://www.forbes.com/profile/brian-chesky/.

91   *At the conclusion of Chesky's creation*: Steven T. Jones, "Forum Begins to Bridge the Housing-Transportation Divide," *San Francisco Bay Guardian*, October 10, 2014, http://www.sfbg.com/politics/2013/04/24/hype-reality-and-accountability-collaborative-consumption.

92   *At last measure, the estimated size*: Sarah Cannon and Lawrence H. Summers, "How Uber and the Sharing Economy Can Win Over Regulators," *Harvard Business Review*, October 13, 2014, https://hbr.org/2014/10/how-uber-and-the-sharing-economy-can-win-over-regulators/; TX Zhuo, "Airbnb and Uber Are Just the Beginning: What's Next for the Sharing Economy," *Entrepreneur*, March 25, 2015, http://www.entrepreneur.com/article/244192.

92   *Founded in 2009 by Travis Kalanick*: Cities, Uber, https://www.uber.com/cities.

92   *Uber's first tagline was*: Kevin Roose, "Uber Might Be More Valuable Than Facebook Someday. Here's Why," *New York Magazine*, December 6, 2013, http://nymag.com/daily/intelligencer/2013/12/uber-might-be-more-valuable-than-facebook.html.

93   *Uber is developing a ride-sharing*: "The City of the Future: One Million Fewer Cars on the Road," *Uber Newsroom*, October 3, 2014, http://blog.uber .com/city-future.

93   *High-profile investors include*: Brad Stone, "Invasion of the Taxi Snatchers: Uber Leads an Industry's Disruption," *Bloomberg Businessweek*, February 20, 2014, http://www.businessweek.com/articles/2014-02-20/uber-leads-taxi-industry-disruption-amid-fight-for-riders-drivers.

95   *Where a typical tourist stay*: "The Economic Impacts of Home Sharing in Cities around the World," Airbnb, https://www.airbnb.com/economic-impact/.

95   *I think it is no coincidence*: Ibid.

96   *As many households faced*: Ibid.

96   *Today you can buy any*: "About Ferrari," eBay, http://www.ebay.com/motors/carsandtrucks/Ferrari.

98   *As of the 2014 holiday season*: Jillian Kumagai, "More Than 21,000 Retailers Accept Bitcoin Payments," *Mashable*, November 15, 2014, http://mashable .com/2014/11/15/bitcoin-retailers-infographic/?utm_cid=mash-com-Tw-main-link; Jon Matonis, "Top 10 Bitcoin Merchant Sites," *Forbes*, May 24, 2013, http://www.forbes.com/sites/jonmatonis/2013/05/24/top-10-bitcoin-merchant-sites/; Benzinga, "What Companies Accept Bitcoin?" Nasdaq, February 4, 2014, http://www.nasdaq.com/article/what-companies-accept-bitcoin-cm323438; Jonas Chokun, "Who Accepts Bitcoins?" *Bitcoin Values*, http://www.bitcoinvalues.net/who-accepts-bitcoins-payment-companies-stores-take-bitcoins.html.

99   *On October 31, 2008, a research paper*: Benjamin Wallace, "The Rise and Fall of Bitcoin," *Wired*, November 23, 2011, http://www.wired.com/magazine/2011/11/mf_bitcoin/.

99   *It called for the creation*: Satoshi Nakamoto, "Bitcoin: A Peer-to-Peer Electronic Cash System," Bitcoin, November 1, 2008, http://bitcoin.org/bitcoin.pdf.

99   *Banks must be trusted*: Joshua Davis, "The Crypto-Currency," *New Yorker*, October 10, 2011, http://www.newyorker.com/reporting/2011/10/10/111010fa_fact_davis.

103 *The goal is for 21 million*: "How Does Bitcoin Work?" *Economist*, April 11, 2013, http://www.economist.com/blogs/economist-explains/2013/04/economist-explains-how-does-bitcoin-work.

103 *At that point, no more*: Alice Truong, "Top 10 Bitcoin Myths Debunked," *CoinDesk*, June 4, 2013, http://www.coindesk.com/top-10-bitcoin-myths-debunked/.

104 *Bitcoin, as a global payment system*: Marc Andreessen, "Why Bitcoin Matters," *New York Times*, January 21, 2014, http://dealbook.nytimes.com/2014/01/21/why-bitcoin-matters/?_php=true&_type=blogs&_r=0.

105 *If you are wondering why*: "Why I'm Interested in Bitcoin," *CDIXON Blog*, December 31, 2013, http://cdixon.org/2013/12/31/why-im-interested-in-bitcoin/.

106 *Just a couple of days later*: Steven Musil, "Bitcoin Exchange BitFloor Halts Operations, Shuts Down," *CNET*, April 17, 2013, http://www.cnet.com/news/bitcoin-exchange-bitfloor-halts-operations-shuts-down/.

109 *In February 2014, hackers*: Mark Memmott, "Mt. Gox Files for Bankruptcy; Nearly $500M of Bitcoins Lost," NPR, February 28, 2014, http://www.npr.org/blogs/thetwo-way/2014/02/28/283863219/mtgox-files-for-bankruptcy-nearly-500m-of-bitcoins-lost.

109 *This bug allowed hackers*: Danny Bradbury, "What the 'Bitcoin Bug' Means: A Guide to Transaction Malleability," *CoinDesk*, February 12, 2014, http://www.coindesk.com/bitcoin-bug-guide-transaction-malleability/.

109 *This accounts for 386 bitcoins*: "The Troubling Holes in MtGox's Account of How It Lost $600 million in Bitcoins," *MIT Technology Review*, April 4, 2014, http://www.technologyreview.com/view/526161/the-troubling-holes-in-mtgoxs-account-of-how-it-lost-600-million-in-bitcoins/.

109 *More damning evidence emerged*: Robert McMillan, "The Inside Story of Mt. Gox, Bitcoin's $460 Million Disaster," *Wired*, March 3, 2014, http://www.wired.com/2014/03/bitcoin-exchange/.

110 *One Bitcoin enthusiast*: Cyrus Farivar, "Man Has NFC Chips Injected into His Hands to Store Cold Bitcoin Wallet," *Ars Technica*, November 15, 2014, http://arstechnica.com/business/2014/11/man-has-nfc-chips-injected-into-his-hands-to-store-cold-bitcoin-wallet/.

110 *The servers are guarded using*: Xapo—About, https://xapo.com/vault/.

111 *From the right, former Federal Reserve chairman Alan Greenspan*: Jeff

Kearns, "Greenspan Says Bitcoin a Bubble without Intrinsic Currency Value," *Bloomberg*, December 4, 2013, http://www.bloomberg.com/news/2013-12-04/greenspan-says-bitcoin-a-bubble-without-intrinsic-currency-value.html.

111 *They have included*: Paul Krugman, "Bitcoin Is Evil," *New York Times*, December 28, 2013, http://krugman.blogs.nytimes.com/2013/12/28/bitcoin-is-evil/?_php=true&_type=blogs&_r=0; Paul Krugman, "Bits and Barbarism," *New York Times*, December 22, 2013, http://www.nytimes.com/2013/12/23/opinion/krugman-bits-and-barbarism.html; Paul Krugman, "Adam Smith Hates Bitcoin," *New York Times*, April 12, 2013, http://krugman.blogs.nytimes.com/2013/04/12/adam-smith-hates-bitcoin/; Paul Krugman, "The Antisocial Network," *New York Times*, April 14, 2013, http://www.nytimes.com/2013/04/15/opinion/krugman-the-antisocial-network.html.

111 *Krugman writes that the rise*: Krugman, "Bits and Barbarism."

111 *Prominent economist Nouriel Roubini*: Erik Holm, "Nouriel Roubini: Bitcoin Is a 'Ponzi Game,'" *Wall Street Journal*, March 10, 2014, http://blogs.wsj.com/moneybeat/2014/03/10/nouriel-roubini-bitcoin-is-a-ponzi-game/.

112 *He finished by going for the economic jugular*: Ibid.

113 *He is on the boards*: "Lending Club Names Lawrence H. Summers to Board of Directors," Lending Club, December 13, 2012, https://www.lendingclub.com/public/lending-club-press-2012-12-13.action; "What We Do," Lending Club, https://www.lendingclub.com/public/about-us.action.

113 *He is also an advisor*: "Investing: Backing Brilliant Entrepreneurs to Build the Future," Andreessen Horowitz, http://a16z.com/team/.

113 *He has even joined*: "Announcing Xapo's Advisory Board," Xapo, May 26, 2015, https://blog.xapo.com/announcing-xapos-advisory-board/.

113 *In April 2015*: Nathaniel Popper, "Goldman and IDG Put $50 Million to Work in a Bitcoin Company," *New York Times*, April 30, 2015, http://www.nytimes.com/2015/04/30/business/dealbook/goldman-and-idg-put-50-million-to-work-in-a-bitcoin-company.html?_r=0.

113 *In December 2013*: Neil Gough, "Bitcoin Value Sinks after Chinese Exchange Move," *New York Times*, December 18, 2013, http://www.nytimes.com/2013/12/19/business/international/china-bitcoin-exchange-ends-renminbi-deposits.html.

113 *In March 2014, the Internal Revenue Service*: Rachel Abrams, "I.R.S. Takes a Position on Bitcoin: It's Property," *New York Times*, March 25, 2014, http://dealbook.nytimes.com/2014/03/25/i-r-s-says-bitcoin-should-be-considered-property-not-currency/.

114 *Yet just three months later*: Byron Tau, "FEC OKs Bitcoin Campaign Donations," *Politico*, May 8, 2014, http://www.politico.com/story/2014/05/fec-oks-bitcoin-campaign-donations-106492.html.

114 *In fact, the government of Canada*: "MintChip—The Evolution of Currency," MintChip Developer Resources, http://developer.mintchipchallenge.com/index.php; Pete Rizzo, "Canadian Government to End 'MintChip' Digital Currency Program," *CoinDesk*, April 4, 2014, http://www.coindesk.com/

canadian-government-end-mintchip-digital-currency-program/; David George-Cosh, "Canada Puts Halt to MintChip Plans; Could Sell Digital Currency Program," *Wall Street Journal*, April 4, 2014, http://blogs.wsj.com/canada-realtime/2014/04/04/canada-puts-halt-to-mintchip-plans-could-sell-digital-currency-program/.

114 *As a result, Bitcoin*: Andreessen, "Why Bitcoin Matters."

114 *As Andreessen further describes*: Brian Fung, "Marc Andreessen: In 20 Years, We'll Talk about Bitcoin Like We Talk about the Internet Today," *Washington Post*, May 21, 2014, http://www.washingtonpost.com/blogs/the-switch/wp/2014/05/21/marc-andreessen-in-20-years-well-talk-about-bitcoin-like-we-talk-about-the-internet-today/.

115 *The great innovation of Tim Berners-Lee*: "Inventor of the Week Archive: The World Wide Web," MIT, http://web.mit.edu/invent/iow/berners-lee.html.

116 *My hunch is that*: Joichi Ito, "Why Bitcoin Is and Isn't like the Internet," *LinkedIn Pulse*, January 18, 2015, https://www.linkedin.com/pulse/why-bitcoin-isnt-like-internet-joichi-ito.

116 *As Marc Andreessen has described*: Andreessen, "Why Bitcoin Matters."

117 *There are now hundreds*: "Crypto-Currency Market Capitalizations," https://coinmarketcap.com/.

118 *Collectively, miners spend nearly*: Mark Gimein, "Virtual Bitcoin Mining Is a Real-World Environmental Disaster," *Bloomberg*, April 12, 2013, http://www.bloomberg.com/news/2013-04-12/virtual-bitcoin-mining-is-a-real-world-environmental-disaster.html; Tim Worstall, "Fascinating Number: Bitcoin Mining Uses $15 Million's Worth of Electricity Every Day," *Forbes*, December 3, 2013, http://www.forbes.com/sites/timworstall/2013/12/03/fascinating-number-bitcoin-mining-uses-15-millions-worth-of-electricity-every-day/.

118 *In 2013, the Bitcoin community*: Michael Carney, "Bitcoin Has a Dark Side: Its Carbon Footprint," *Pando*, December 16, 2013, http://pando.com/2013/12/16/bitcoin-has-a-dark-side-its-carbon-footprint/.

118 *A British programmer decided*: Nathaniel Popper, "Into the Bitcoin Mines," *New York Times*, December 21, 2013, http://dealbook.nytimes.com/2013/12/21/into-the-bitcoin-mines/.

118 *Charlie Lee, a former Google*: Danny Bradbury, "Litecoin Founder Charles Lee on the Origins and Potential of the World's Second Largest Cryptocurrency," *CoinDesk*, July 23, 2013, http://www.coindesk.com/litecoin-founder-charles-lee-on-the-origins-and-potential-of-the-worlds-second-largest-cryptocurrency/.

118 *"People like choices"*: Nathaniel Popper, "In Bitcoin's Orbit: Rival Virtual Currencies Vie for Acceptance," *New York Times*, November 24, 2013, http://dealbook.nytimes.com/2013/11/24/in-bitcoins-orbit-rival-virtual-currencies-vie-for-acceptance/.

118 *he designed the Litecoin*: Robert McMillan, "Ex-Googler Gives the World a Better Bitcoin," *Wired*, August 30, 2013, http://www.wired.com/2013/08/litecoin/.

118 *Lee also chose this*: "What Is the Difference between Litecoin and Bitcoin?" *CoinDesk*, April 2, 2014, http://www.coindesk.com/information/comparing-litecoin-bitcoin/.

118 *Ripple markets itself as*: Ripple, https://ripple.com/.

119 *It has been compared to a* hawala: "Hawala and Alternative Remittance Systems," US Department of Treasury: Resource Center, http://www.treasury.gov/resource-center/terrorist-illicit-finance/Pages/Hawala-and-Alternatives.aspx; Antony Lewis, "Ripple Explained: Medieval Banking with a Digital Twist," *CoinDesk*, May 11, 2014, http://www.coindesk.com/ripple-medieval-banking-digital-twist/.

119 *Ripple Labs maintains the global ledger*: Brad Stone, "Introducing Ripple, a Bitcoin Copycat," *Bloomberg Businessweek*, April 11, 2013, http://www.bloomberg.com/bw/articles/2013-04-11/introducing-ripple-a-bitcoin-copycat; Bryant Gehring, "How Ripple Works," Ripple, October 16, 2014, https://ripple.com/knowledge_center/how-ripple-works/; "XRP Distribution," Ripple Labs, https://www.ripplelabs.com/xrp-distribution/.

119 *The rest will be*: "XRP Distribution."

119 *Ripple is backed by*: Stone, "Introducing Ripple."

CHAPTER 4: THE WEAPONIZATION OF CODE

121 *On Wednesday, August 15, 2012*: Nicole Perlroth, "In Cyberattack on Saudi Firm, US Sees Iran Firing Back," *New York Times*, October 23, 2012, http://www.nytimes.com/2012/10/24/business/global/cyberattack-on-saudi-oil-firm-disquiets-us.html; "Saudi Aramco: 12.5 million barrels per day," *Forbes*, http://www.forbes.com/pictures/mef45glfe/1-saudi-aramco-12-5-million-barrels-per-day-3/.

121 *It infected not just Saudi Aramco's*: Christopher Bronk and Eneken Tikk-Ringas, "Hack or Attack? Shamoon and the Evolution of Cyber Conflict," *Survival, Global Politics and Strategy*, February 1, 2013, http://bakerinstitute.org/files/641/.

122 *In order to permanently delete*: Chris Bronk, interview with Jennifer Citak, December 20, 2013.

122 *For good measure, Shamoon also*: "The Shamoon Attacks," *Symantec Blog*, August 16, 2012, http://www.symantec.com/connect/blogs/shamoon-attacks.

122 *Shamoon went beyond wiping out*: Nicole Perlroth, "Connecting the Dots after Cyberattack on Saudi Aramco," *New York Times*, August 27, 2012, http://bits.blogs.nytimes.com/2012/08/27/connecting-the-dots-after-cyberattack-on-saudi-aramco/.

122 *The attacker was sent the IP address*: "The Shamoon Attacks."

122 *The virus was discovered*: Ibid.; "Shamoon the Wiper: Copycats at Work," *Securelist*, August 16, 2012, http://www.securelist.com/en/blog/208193786/Shamoon_the_Wiper_Copycats_at_Work; Aviv Raff, "Shamoon, a Two-Stage Targeted Attack," *Seculert*, August 2012, http://www.seculert.com/blog/2012/08/shamoon-two-stage-targeted-attack.html.

122 *It took two weeks*: Bronk, interview.

122 *By the time the attack*: Ibid.; "The Shamoon Attacks."

122 *Two weeks later, Shamoon*: Bronk, interview; Camilla Hall and Javier Blas, "Aramco Cyber Attack Targeted Production," *Financial Times*, December 10, 2012, http://www.ft.com/intl/cms/s/0/5f313ab6-42da-11e2-a4e4-00144feabdc0 .html#axzz2qP9F3kEY; Bronk and Tikk-Ringas, "Hack or Attack?"

122 *Saudi Aramco is responsible*: Parag Khanna, "The Rise of Hybrid Governance," McKinsey & Company, October 2012, http://www.mckinsey.com/ insights/public_sector/the_rise_of_hybrid_governance.

123 *Little did it know that its*: Christopher Bronk and Eneken Tikk-Ringas, "The Cyber Attack on Saudi Aramco," *Survival: Global Politics and Strategy*, April–May 2013, 81–96, http://www.iiss.org/en/publications/ survival/sections/2013-94b0/survival–global-politics-and-strategy-april- may-2013-b2cc/55-2-08-bronk-and-tikk-ringas-e272; Jim Garamone, "Panetta Spells Out DOD Roles in Cyberdefense," American Forces Press Service, US Department of Defense, October 11, 2012, http://www.defense.gov/news/ newsarticle.aspx?id=118187.

123 *Saudi Aramco is the most valuable*: Charles Orton-Jones, "Stop the Press! Apple Is NOT the World's Most Valuable Company," *London- lovesBusiness*, August 21, 2012, http://www.londonlovesbusiness.com/ business-news/finance/stop-the-press-apple-is-not-the-worlds-most- valuable-company/3250.article.

123 *If a cyberattack could happen*: Bronk and Tikk-Ringas, "Hack or Attack?"

123 *It is perhaps ironic that*: "Paul Baran and the Origins of the Internet," Rand Corporation, http://www.rand.org/about/history/baran.list.html.

124 *Whether motivated by politics*: "Net Losses: Estimating the Global Cost of Cybercrime: Economic Impact of Cybercrime II," Center for Strategic and International Studies, June 2014, http://www.mcafee.com/us/resources/reports/ rp-economic-impact-cybercrime2.pdf.

125 *The message read*: Paul Marks, "Dot-Dash-Diss: The Gentleman Hacker's 1903 Lulz," *New Scientist*, December 20, 2011, https://www.newscientist.com/ article/mg21228440-700-dot-dash-diss-the-gentleman-hackers-1903-lulz/.

125 *There are three main types*: Peter W. Singer and Allan Friedman, *Cybersecurity: What Everyone Needs to Know* (New York: Oxford University Press, 2014), 69; Bronk, interview.

125 *Attacks that compromise confidentiality*: Bronk, interview.

125 *By inserting malware—malicious software*: Jennifer Bjorhus, "A Year Later, No Charges for Target Hack," *Portland Press Herald*, November 25, 2014, http://www.pressherald.com/2014/11/25/a-year-later-no-charges-for-target- hack/.

125 *In addition, the hackers stole*: Mark Hosenball, "Target Vendor Says Hackers Breached Data Link Used for Billing," Reuters, February 6, 2014, http:// www.reuters.com/article/2014/02/06/us-target-breach-vendor-idUS- BREA1523E20140206.

125   *Profits fell 46 percent in*: Elizabeth A. Harris, "Faltering Target Parts Ways with Chief," *New York Times*, May 6, 2014, http://www.nytimes.com/2014/05/06/business/target-chief-executive-resigns.html?ref=technology&_r=0.

125   *The company could still face*: Brian Krebs, "Target Hackers Broke in via HVAC Company," *Krebs on Security* (blog), February 5, 2014, http://krebsonsecurity.com/2014/02/target-hackers-broke-in-via-hvac-company/.

125   *It lost billions of dollars*: Susan Taylor, Siddharth Cavale, and Jim Finkle, "Target's Decision to Remove CEO Rattles Investors," Reuters, May 5, 2014, http://www.reuters.com/article/2014/05/05/us-target-ceo-idUSBREA440BD20140505.

126   *These types of attacks*: Vangie Beal, "DDoS attack: Distributed Denial of Service," Webopedia, http://www.webopedia.com/TERM/D/DDoS_attack.html.

126   *One of the largest cyberattacks*: Parmy Olson, "The Largest Cyber Attack in History Has Been Hitting Hong Kong Sites," *Forbes*, November 20, 2014, http://www.forbes.com/sites/parmyolson/2014/11/20/the-largest-cyber-attack-in-history-has-been-hitting-hong-kong-sites/.

126   *Last, cyberattacks can also*: Bronk, interview.

126   *Just after 1:00 p.m., the AP's Twitter*: Max Fisher, "Syrian Hackers Claim AP Hack That Tipped Stock Market by $136 billion. Is It Terrorism?" *Washington Post*, April 23, 2013, http://www.washingtonpost.com/blogs/worldviews/wp/2013/04/23/syrian-hackers-claim-ap-hack-that-tipped-stock-market-by-136-billion-is-it-terrorism/.

127   *The term* phishing *originated from*: "Phishing," Language Log, University of Pennsylvania, http://itre.cis.upenn.edu/~myl/languagelog/archives/001477.html.

127   *When the staffer clicked on*: Fisher, "Syrian Hackers Claim AP Hack."

127   *While I was at the State Department in 2011*: "War in the Fifth Domain," *Economist*, July 1, 2010, http://www.economist.com/node/16478792.

128   *For example, when the United States was planning*: Ellen Nakashima, "US Cyberweapons Had Been Considered to Disrupt Gaddafi's Air Defenses," *Washington Post*, October 17, 2011, http://www.washingtonpost.com/world/national-security/us-cyber-weapons-had-been-considered-to-disrupt-gaddafis-air-defenses/2011/10/17/gIQAETpssL_story.html.

128   *In 2002, the Dalai Lama*: Desmond Ball, "China's Cyber Warfare Capabilities," *Security Challenges* 7, no. 2 (2011): 81–103, http://www.securitychallenges.org.au/ArticlePDFs/vol7no2Ball.pdf.

129   *An American group of powerhouse*: "The IP Commission Report," Commission on the Theft of American Intellectual Property by the National Bureau of Asian Research, May 2013, http://www.ipcommission.org/report/ip_commission_report_052213.pdf.

129   *The NSA director at the time*: Keith B. Alexander, "Cybersecurity and American Power: Addressing New Threats to America's Economy and Military" (presentation at the American Enterprise Institute, Washington, DC, July 9, 2012).

129 *The Canadian telecommunications company*: Amanda Hoyle, "Nortel Fell Hard, and Only 20 Workers Are Still Here," *Triangle Business Journal*, January 11, 2013, http://www.bizjournals.com/triangle/print-edition/2013/01/11/nortel-fell-hard-and-only-20-workers.html?page=all.

129 *From 2000 to 2009, when Nortel Networks*: Gerry Smith, "Hackers Cost US Economy Up to 500,000 Jobs Each Year, Study Finds," *Huffington Post*, July 25, 2013, http://www.huffingtonpost.com/2013/07/25/hackers-jobs_n_3652893.html; "Nortel Collapse Linked to Chinese Hackers," *CBC News*, February 15, 2012, http://www.cbc.ca/news/business/nortel-collapse-linked-to-chinese-hackers-1.1260591.

129 *In 2013, an American cybersecurity company*: "APT1: Exposing One of China's Cyber Espionage Units," *Mandiant*, February 18, 2013, http://intelreport.mandiant.com/Mandiant_APT1_Report.pdf.

129 *Working out of a location*: Zoe Li, "What We Know about the Chinese Army's Alleged Cyber Spying Unit," CNN, May 20, 2014, http://www.cnn.com/2014/05/20/world/asia/china-unit-61398/.

129 *Since 2006, the unit*: "China vs US, Cyber Superpowers Compared," InfoSec Institute, June 10, 2013, http://resources.infosecinstitute.com/china-vs-us-cyber-superpowers-compared/.

130 *The indictment alleges*: "US Charges Five Chinese Military Hackers for Cyber Espionage against US Corporations and a Labor Organization for Commercial Advantage," *Justice News*, US Department of Justice, May 19, 2014, http://www.justice.gov/opa/pr/us-charges-five-chinese-military-hackers-cyber-espionage-against-us-corporations-and-labor.

130 *"The Chinese government and Chinese military"*: Jack Gillum and Eric Tucker, "US Hacking Victims Fell Prey to Mundane Ruses," Associated Press, May 20, 2014, http://bigstory.ap.org/article/us-hacking-victims-fell-prey-mundane-ruses.

131 *A spokesman for North Korea's*: Brooks Barnes and Nicole Perlroth, "Sony Films Are Pirated, and Hackers Leak Studio Salaries," *New York Times*, December 2, 2014, http://www.nytimes.com/2014/12/03/business/media/sony-is-again-target-of-hackers.html?_r=0.

131 *Notably, all of North Korea's Internet*: Jack Kim and Lesley Wroughton, "North Korea's Internet Links Restored amid US Hacking Dispute," Reuters, December 23, 2014, http://www.reuters.com/article/2014/12/23/us-north-korea-cyberattack-idUSKBN0K107920141223.

132 *Cisco Systems chairman John Chambers*: "Cisco Keynote Highlights from CES 2014," YouTube, January 10, 2014, http://www.youtube.com/watch?v=TepUznT42ro.

132 *From 2015 to 2020, the number*: "The Internet of Things Will Drive Wireless Connected Devices to 40.9 Billion in 2020," ABI Research, August 20, 2014, https://www.abiresearch.com/press/the-internet-of-things-will-drive-wireless-connect.

132 *Chambers predicts that the Internet of Things*: Don Clark, "Cisco CEO

Chambers Still Biggest 'Internet of Things' Cheerleader," *Wall Street Journal*, January 7, 2014, http://blogs.wsj.com/digits/2014/01/07/cisco-ceo-john-chambers-Internet-of-everything-ces-2014/.

132   *For context, the GDP*: "Report for Selected Country Groups and Subjects," International Monetary Fund: World Economic Outlook Database, October 2014, http://www.imf.org/external/pubs/ft/weo/2014/02/weodata/weorept .aspx?pr.x=41&pr.y=10&sy=2014&ey=2014&scsm=1&ssd=1&sort=country& ds=.&br=1&c=001%2C998&s=NGDPD%2CPPPGDP&grp=1&a=1.

132   *The first is the number of*: Keith Naughton, "The Race to Market the Connected Car," *Automotive News*, January 10, 2014, http://www .autonews.com/article/20140110/OEM06/301109910/the-race-to-market-the-connected-car.

132   *According to a Juniper research report*: "Smart Home Revenues to Reach $71 Billion by 2018, Juniper Research Finds," Juniper Research, February 11, 2014, http://www.juniperresearch.com/viewpressrelease.php?pr=429.

132   *A McKinsey report projects*: James Manyika and Michael Chui, "All Things Online," McKinsey Global Institute, McKinsey & Company, September 23, 2013, http://www.mckinsey.com/insights/mgi/in_the_news/ all_things_online.

133   *In the Target hack*: Krebs, "Target Hackers Broke in Via HVAC Company."

133   *Target is a company with*: "Corporate Fact Sheet," Target, http://pressroom .target.com/corporate.

134   *In January 2014, security provider*: "Proofpoint Uncovers Internet of Things (IoT) Cyberattack," *Proofpoint*, January 16, 2014, http://www.proofpoint .com/about-us/press-releases/01162014.php.

134   *Mikko Hypponen, a Finnish cybersecurity*: Mikko Hypponen, interview with Jennifer Citak, October 22, 2013.

136   *He has also earned*: "Legion of Merit," MyServicePride.com, http://www .myservicepride.com/content/legion-of-merit/.

136   *In the lobby of the Marriott Key Bridge*: Jim Gosler, interview with Ari Ratner, March 25, 2014.

137   *"There are about one thousand security people"*: "The United States Cyber Challenge," White House, SANS Institute, May 8, 2009, http://www .whitehouse.gov/files/documents/cyber/The%20United%20States%20 Cyber%20Challenge%201.1%20(updated%205-8-09).pdf.

137   *Cybersecurity professionals in the United States*: Kenneth Corbin, "Cybersecurity Pros in High Demand, Highly Paid and Highly Selective," CIO, August 8, 2013, http://www.cio.com/article/2383451/careers-staffing/ cybersecurity-pros-in-high-demand—highly-paid-and-highly-selective .html.

137   *Gosler's is not a lone voice*: "Remarks as delivered by The Honorable James R. Clapper, Director of National Intelligence, Opening Statement to the Worldwide Threat Assessment Hearing Senate Armed Services Committee," Office of the Director of National Intelligence, February 26, 2015, http://www.dni

.gov/files/documents/2015%20WWTA%20As%20Delivered%20DNI%20 Oral%20Statement.pdf.

138 *The malware was "designed*: "Snake Campaign & Cyber Espionage Toolkit," BAE Systems Applied Intelligence, 2014, http://info.baesystemsdetica.com/ rs/baesystems/images/snake_whitepaper.pdf.

139 *Evidence, such as the time zone*: David E. Sanger and Steven Erlanger, "Suspicion Falls on Russia as 'Snake' Cyberattacks Target Ukraine's Government," *New York Times*, March 8, 2014, http://www.nytimes.com/2014/03/09/world/ europe/suspicion-falls-on-russia-as-snake-cyberattacks-target-ukraines- government.html.

139 *FireEye, a global network security*: Pierluigi Paganini, "Russia and Ukraine Cyber Dispute Analyzed by FireEye," *Security Affairs*, May 30, 2014, http:// securityaffairs.co/wordpress/25369/intelligence/russia-and-ukraine-cyber- tension.html.

139 *One pro-Russia member of Parliament justified*: "M. Saakashvili Is Denied to Enter into Ukraine," *Times.am*, December 24, 2013, http://times .am/?p=36719&l=en.

140 *According to SBU head Valentyn Nalyvaichenko*: "Ukraine: Electoral Committee Cyber-Virus 'Liquidated'—SBU Chief," YouTube, May 23, 2014, https://www.youtube.com/watch?v=u354nFMRv1Q.

140 *As SBU was foiling the attack*: "Russian TV Announces Right Sector Leader Led Ukraine Polls," Radio Free Europe/Radio Liberty, May 26, 2014, http:// www.rferl.org/content/russian-tv-announces-right-sector-leader-yarosh-led- ukraine-polls/25398882.html.

140 *The screenshot was from the hacked site*: "Security Service of Ukraine Ensured Protection and Safe Functioning of Telecommunication System of the Central Electoral Commission during Elections of the President of Ukraine," Security Service of Ukraine, May 27, 2014, http://www.sbu.gov.ua/sbu/ control/en/publish/article?art_id=126126&cat_id=35317&mustWords=discr edit&searchPublishing=1.

140 *Pro-Russian hacktivist group CyberBerkut*: CyberBerkut, http://www.cyber- berkut.org/.

141 *The wave of denial-of-service attacks*: Jeremy Hsu, "Why There's No Real Cyber- war in the Ukraine Conflict," *IEEE Spectrum*, March 14, 2014, http://spectrum .ieee.org/tech-talk/computing/networks/why-theres-no-real-cyberwar-in-the- ukraine-conflict.

141 *Eventually a Kremlin-backed patriotic*: Charles Clover, "Kremlin-Backed Group behind Estonia Cyber Blitz," *Financial Times*, March 11, 2009, http://www.ft.com/intl/cms/s/0/57536d5a-0ddc-11de-8ea3-0000779fd2ac .html#axzz33Us0YfDw.

141 *A year after the Nashi cyberattacks*: John Markoff, "Before the Gunfire, Cyberattacks," *New York Times*, August 12, 2008, http://www.nytimes .com/2008/08/13/technology/13cyber.html.

141 *The website of the National Bank of Georgia*: Eneken Tikk, Kadri Kaska,

Kristel Rünnimeri, et al., *Cyber Attacks Against Georgia: Legal Lessons Identified* (Tallinn, Estonia: Cooperative Cyber Defense Center of Excellence, 2008), http://www.carlisle.army.mil/DIME/documents/Georgia%201%200.pdf.

143 *One of the most notable cyberattacks*: Ariana Eunjung Cha and Ellen Nakashima, "Google China Cyberattack Part of Vast Espionage Campaign, Experts Say," *Washington Post*, January 14, 2010, http://www.washingtonpost.com/wp-dyn/content/article/2010/01/13/AR2010011300359.html.

144 *Throughout summer and fall*: Michael Corkery, Jessica Silver-Greenberg, and David E. Sanger, "Obama Had Security Fears on JPMorgan Data Breach," *New York Times*, October 8, 2014, http://dealbook.nytimes.com/2014/10/08/cyberattack-on-jpmorgan-raises-alarms-at-white-house-and-on-wall-street/?_r=0.

144 *Items of interest will be located*: Spencer Ackerman, "CIA Chief: We'll Spy on You through Your Dishwasher," *Wired*, March 15, 2012, http://www.wired.com/2012/03/petraeus-tv-remote.

145 *One study pegged the lost business*: Daniel Castro, "How Much Will PRISM Cost the US Cloud Computing Industry?" Information Technology and Innovation Foundation, August 2013, http://www2.itif.org/2013-cloud-computing-costs.pdf.

145 *In February 2015, President Obama*: "Executive Order: Improving Critical Infrastructure Cybersecurity," White House, February 12, 2013, https://www.whitehouse.gov/the-press-office/2013/02/12/executive-order-improving-critical-infrastructure-cybersecurity.

146 *Justifying the unusual step*: Joyce Brayboy, "Army Cyber Defenders Open Source Code in new GitHub Project," US Army, January 28, 2015, http://www.army.mil/article/141734/Army_cyber_defenders_open_source_code_in_new_GitHub_project/.

147 *A dozen years ago*: Michael Peck, "Cybersecurity Market to Hit $77B," *Federal Times*, February 21, 2014, http://www.federaltimes.com/article/20140221/CYBER/302210004/Cybersecurity-market-hit-77B; Fahmida Y. Rashid, "Global Cybersecurity Market to Hit $120.1 Billion by 2017," *Security Current*, March 6, 2014, http://www.securitycurrent.com/en/news/ac_news/global-cybersecurity-marke.

147 *"Spending, globally, continues to remain"*: Peck, "Cybersecurity Market to Hit $77B."

147 *"I think it will continue"*: Peter Singer, interview with Jennifer Citak, October 22, 2013; P. W. Singer and Allan Friedman, *Cybersecurity and Cyberwar: What Everyone Needs to Know* (Oxford, UK: Oxford University Press, 2014).

147 *Finnish cybersecurity expert Mikko Hypponen*: Hypponen, interview with Citak.

CHAPTER 5: DATA: THE RAW MATERIAL OF THE INFORMATION AGE

153 *Private companies now collect*: Josh Gerstein and Stephanie Simon, "Who Watches the Watchers? Big Data Goes Unchecked," *Politico*, May 14, 2014, http://www.politico.com/story/2014/05/big-data-beyond-the-nsa-106653.html.

153 *In the first 50 years*: Elizabeth L. Eisenstein, *The Printing Revolution in Early Modern Europe* (Cambridge, UK: Cambridge University Press, 2005).

154 *By 1996, there was so*: "The Evolution of Storage Systems," *IBM Systems Journal* 42, no. 2 (2003): 205–17, http://ieeexplore.ieee.org/xpl/login.jsp?re load=true&tp=&arnumber=5386860&url=http%3A%2F%2Fieeexplore.ieee .org%2Fstamp%2Fstamp.jsp%3Ftp%3D%26arnumber%3D5386860.

154 *Less than a decade later*: James Manyika, Michael Chui, Brad Brown, et al., "Big Data: The Next Frontier for Innovation, Competition, and Productivity," McKinsey Global Institute, May 2011, http://www.mckinsey.com/insights/ business_technology/big_data_the_next_frontier_for_innovation.

154 *Ninety percent of the world's digital data*: "Big Data, for Better or Worse: 90% of World's Data Generated over Last Two Years," *Science Daily*, May 22, 2013, http://www.sciencedaily.com/releases/2013/05/130522085217.htm.

154 *Every year, the amount of*: Steve Lohr, "The Age of Big Data," *New York Times*, February 11, 2012, http://www.nytimes.com/2012/02/12/sunday-review/big-datas-impact-in-the-world.html?pagewanted=all.

154 *Every minute of the day*: "Data Never Sleeps 2.0," Domo, http://www.domo .com/learn/data-never-sleeps-2.

154 *The sum of all this*: Patrick Tucker, "Has Big Data Made Anonymity Impossible?" *MIT Technology Review*, May 7, 2013, http://www.technologyreview .com/news/514351/has-big-data-made-anonymity-impossible/.

155 *In one instance, the campaign*: "Inside the Cave: An In-Depth Look at the Digital, Technology, and Analytics Operations of Obama for America," Engage Research, 2012, http://enga.ge/download/Inside%20the%20Cave.pdf.

155 *As a senior member of*: Ibid.

155 *In 2012, the Obama campaign*: Kenneth P. Vogel, Dave Levinthal, and Tarini Parti, "Barack Obama, Mitt Romney Both Topped $1 Billion in 2012," *Politico*, December 7, 2012, http://www.politico.com/story/2012/12/barack-obama-mitt-romney-both-topped-1-billion-in-2012-84737.html; "Inside the Cave: An In-Depth Look at the Digital, Technology, and Analytics Operations of Obama for America," Engage Research, 2012, http://enga.ge/download/ Inside%20the%20Cave.pdf.

155 *The Obama operation was about*: Zac Moffatt, "Successes of the Romney and Republican Digital Efforts in 2012," Targeted Victory, December 11, 2012, http:// www.targetedvictory.com/2012/12/11/success-of-the-romney-republican-digital-efforts-2012/; "Inside the Cave."

156 *And my dad looks at me*: Dan Wagner, interview with Ari Ratner, May 28, 2014.

156 *And you can do that now*: Ibid.

156 *Typically large data analysis*: Michael Slaby, interview with Ari Ratner, December 2, 2013.

157 *Big data's really just*: Ibid.

157 *It examines small facts and aggregates*: Erik Brynjolfsson and Andrew McAfee, "The Big Data Boom Is the Innovation Story of Our Time," *Atlantic*, November

21, 2011, http://www.theatlantic.com/business/archive/2011/11/the-big-data-boom-is-the-innovation-story-of-our-time/248215/; Zeynep Tufekci, "Engineering the Public: Big Data, Surveillance and Computational Politics," *First Monday* 19, no. 7 (2014), http://firstmonday.org/ojs/index.php/fm/article/view/4901/4097.

158 *Today's translation tools*: Stephen Shankland, "Google Translate Now Serves 200 Million People Daily," *CNET*, May 18, 2013, http://www.cnet.com/news/google-translate-now-serves-200-million-people-daily/.

160 *If one does not need to be*: Ethnologue: Languages of the World, http://www.ethnologue.com/.

160 *Papua New Guinea is loaded*: "Papua New Guinea's Fisheries Boom," Business Advantage PNG, February 27, 2013, http://www.businessadvantagepng.com/papua-new-guineas-fisheries-boom/.

161 *The World Food Programme reports*: "Hunger Statistics," World Food Programme, http://www.wfp.org/hunger/stats.

161 *This comes in the midst of*: Ulisses Mello and Lloyd Treinish, "Precision Agriculture: Using Predictive Weather Analytics to Feed Future Generations," IBM Research, http://www.research.ibm.com/articles/precision_agriculture.shtml.

163 *It figured out that analytics*: Rob Thomas and Patrick McSharry, *Big Data Revolution: What Farmers, Doctors and Insurance Agents Teach Us about Discovering Big Data Patterns* (Hoboken, NJ: Wiley, 2015).

164 *In 2014 Monsanto's chief technology officer*: Howard Baldwin, "Big Data Hits the Dirt," *Forbes*, December 8, 2014, http://www.forbes.com/sites/howard-baldwin/2014/12/08/big-data-hits-the-dirt/.

165 *Farming is so difficult in India*: P. Sainath, "Have India's Farm Suicides Really Declined?" *BBC News*, July 14, 2014, http://www.bbc.com/news/world-asia-india-28205741.

165 *What's more, 25 percent of the*: "Number of Hungry People in India Falling: UN Report," *Zee News*, September 18, 2014, http://zeenews.india.com/business/news/economy/number-of-hungry-people-in-india-falling-un-report_108529.html.

166 *Nitrogen from fertilizer creates*: Tom Laskawy, "New Science Reveals Agriculture's True Climate Impact," *Grist*, April 10, 2012, http://grist.org/climate-change/new-science-reveals-agricultures-true-climate-impact/.

166 *Fertilizer produces nitrous oxide*: S. Park, P. Croteau, K. A. Boering, et al., "Trends and Seasonal Cycles in the Isotopic Composition of Nitrous Oxide since 1940," *Nature Geoscience* 5 (2012): 261–65, http://www.nature.com/ngeo/journal/v5/n4/full/ngeo1421.html.

166 *Of the roughly 7 billion shares*: Viktor Mayer-Schönberger and Kenneth Cukier, *Big Data: A Revolution That Will Transform How We Live, Work, and Think* (London: John Murray Publishers, 2013).

167 *In 2013, these firms collectively*: Lisa Fleisher, "London's Financial Tech Sector Growth Outpaces Global Growth," *Wall Street Journal*, March 26,

2014, http://blogs.wsj.com/digits/2014/03/26/londons-financial-tech-sector-growth-fastest-in-world/; Laura Lorenzetti, "Big Banks Are Shunning Tradition and Turning to Tech Startups," *Fortune*, June 26, 2014, http://fortune .com/2014/06/26/big-banks-are-shunning-tradition-and-turning-to-tech-startups/.

167   *"We owe it to ourselves"*: Lorenzetti, "Big Banks Are Shunning Tradition."

169   *Faced with regulatory hurdles*: Standard Treasury, https://standardtreasury .com/; "Standard Treasury Joins Silicon Valley Bank," Zac Townsend blog, http://blog.zactownsend.com/standard-treasury-joins-silicon-valley-bank.

171   *No one else has*: Mike Isaac, "Square Expands Its Reach into Small-Business Services," *New York Times*, March 8, 2015, http://www.nytimes .com/2015/03/09/technology/the-payment-start-up-square-expands-its-reach-into-small-businesses.html.

172   *The captain and his colleagues*: Brendan McGarry, "Special Forces, Marines Embrace Palantir Software," *Defense Tech*, July 1, 2013, http://defensetech .org/2013/07/01/special-forces-marines-embrace-palantir-software/; "Better Measures and Plans Needed to Help Achieve Enterprise Intelligence Sharing Goals," Government Accountability Office, June 2013, http://images.military .com/PDF/gao-report-dcgs-063013.pdf; Andy Greenberg, "How a 'Deviant' Philosopher Built Palantir, a CIA-Funded Data-Mining Juggernaut," *Forbes*, August 14, 2013, http://www.forbes.com/sites/andygreenberg/2013/08/14/ agent-of-intelligence-how-a-deviant-philosopher-built-palantir-a-cia-funded-data-mining-juggernaut/.

172   *The company is run*: Jon Xavier, "Four Things We Learned about Palantir CEO Alex Karp," *Silicon Valley Business Journal*, August 19, 2013, http://www.bizjournals.com/sanjose/news/2013/08/15/palantirs-alex-karp .html?page=all.

172   *Karp was a student under*: James Bohman and William Rehg, "Jürgen Habermas," *Stanford Encyclopedia of Philosophy*, August 4, 2014, http://plato .stanford.edu/entries/habermas/#ImpTraWor.

173   *Its technology has been used*: Greenberg, "How a 'Deviant' Philosopher Built Palantir."

173   *Karp now travels with*: Ibid.

173   *It describes its core disciplines*: Kevin Simler, "Palantir: So What Is It You Guys Do?" Palantir, December 4, 2007, https://www.palantir.com/2007/12/ what-do-we-do/.

173   *An investor prospectus for Palantir*: Matt Burns, "Leaked Palantir Doc Reveals Uses, Specific Functions and Key Clients," *TechCrunch*, January 11, 2015, http://techcrunch.com/2015/01/11/leaked-palantir-doc-reveals-uses-specific-functions-and-key-clients/.

173   *While it is not forced*: Greenberg, "How a 'Deviant' Philosopher Built Palantir."

173   *When recruiting, he says*: Cadie Thompson, "Free Advice: Don't Go Public, Says Palantir's CEO," CNBC, March 19, 2014, http://www.cnbc.com/ id/101507813.

176 *The idea behind the app*: Caitlin Dewey, "The Hot New 'Consent' App, Good2Go, Is Logging the Name and Phone Number of Everyone You Have Sex With," *Washington Post*, September 29, 2014, http://www .washingtonpost.com/news/the-intersect/wp/2014/09/29/the-hot-new-consent-app-good2go-is-logging-the-name-and-phone-number-of-everyone-you-have-sex-with/?Post+generic=%3Ftid%3Dsm_twitter_washingtonpost.

176 *A $100 million database called inBloom*: Stephanie Simon, "Big Brother: Meet the Parent," *Politico*, June 5, 2014, http://www.politico.com/story/2014/06/ internet-data-mining-children-107461.html.

177 *A 2013 Senate Commerce Committee report*: "A Review of the Data Broker Industry: Collection, Use, and Sale of Consumer Data for Marketing Purposes," Office of Oversight and Investigations, Majority Staff, Senate Committee, December 18, 2013, http://www.commerce.senate.gov/public/?a=Files .Serve&File_id=0d2b3642-6221-4888-a631-08f2f255b577.

177 *The lists included domestic violence*: Elizabeth Dwoskin, "Data Broker Removes Rape-Victims List after Journal Inquiry," *Wall Street Journal*, December 19, 2013, http://blogs.wsj.com/digits/2013/12/19/data-broker-removes-rape-victims-list-after-journal-inquiry/; Tara Culp-Ressler, "Big Data Companies Are Selling Lists of Rape Victims to Marketing Firms," *ThinkProgress*, December 19, 2013, http://thinkprogress.org/health/2013/12/19/3089591/big-data-health-data-mining/.

178 *Margo Seltzer, a professor of computer science*: Natasha Lomas, "What Happens to Privacy When the Internet Is in Everything?" *TechCrunch*, January 25, 2015, http://techcrunch.com/2015/01/25/what-happens-to-privacy-when-the-internet-is-in-everything/; AFP and Mark Prigg, "Harvard Professors Warn 'Privacy Is Dead' and Predict Mosquito-Sized Robots That Steal Samples of Your DNA," *Daily Mail*, January 22, 2015, http://www.dailymail .co.uk/sciencetech/article-2921758/Privacy-dead-Harvard-professors-tell-Davos-forum.html#ixzz3PgIkOaR8.

178 *I can erase your address*: Jonathan Shaw, "Why 'Big Data' Is a Big Deal," *Harvard Magazine*, March–April 2014, http://harvardmagazine.com/2014/03/ why-big-data-is-a-big-deal.

181 *Yet we're already ceding it*: John T. Cacioppo, Stephanie Cacioppo, Gian C. Gonzaga et al., "Marital Satisfaction and Break-Ups Differ across On-Line and Off-Line Meeting Venues," *Proceedings of the National Academy of Sciences* 110, no. 25 (2013), http://www.pnas.org/content/110/25/10135.full.

181 *Critics like writer Leon Wieseltier*: Leon Wieseltier, "What Big Data Will Never Explain," *New Republic*, March 26, 2013.

182 *As a response to this*: http://openag.io/about-us/principals-use-cases/.

182 *Data analysis is pretty bad*: David Brooks, "What Data Can't Do," *New York Times*, February 18, 2013, http://www.nytimes.com/2013/02/19/opinion/ brooks-what-data-cant-do.html?_r=0.

183 *When Harvard University's big data*: Kalev Leetaru, "Why Big Data Missed the Early Warning Signs of Ebola," *Foreign Policy*, September 26, 2014, http://foreignpolicy .com/2014/09/26/why-big-data-missed-the-early-warning-signs-of-ebola/#trending.

183 *Once it was clear that Ebola*: "Ebola Cases Could Skyrocket by 2015, Says CDC," Centers for Disease Control and Prevention, *Morbidity and Mortality Weekly Report* 63, *Washington Post*, http://apps.washingtonpost.com/g/page/national/ebola-cases-could-skyrocket-by-2015-says-cdc/1337/.

183 *The actual number ended*: Data Team, "Ebola in Graphics: The Toll of a Tragedy," *Economist*, July 8, 2015, http://www.economist.com/blogs/graphicdetail/2015/02/ebola-graphics.

185 *The best approach to big data*: Slaby, interview.

CHAPTER 6: THE GEOGRAPHY OF FUTURE MARKETS

186 *Marc Andreessen writes*: Marc Andreessen, "Turn Detroit into Drone Valley," *Politico*, June 15, 2014, http://www.politico.com/magazine/story/2014/06/turn-detroit-into-drone-valley-107853.html#ixzz3SwRDqcxw.

189 *One of the world's largest cybersecurity*: Carol Matlack, Michael Riley, and Jordan Robertson, "The Company Securing Your Internet Has Close Ties to Russian Spies," *Bloomberg Businessweek*, March 19, 2015, http://www.bloomberg.com/news/articles/2015-03-19/cybersecurity-kaspersky-has-close-ties-to-russian-spies.

189 *Over the past five years*: "The Boom in Global Fintech Investment," Accenture, 2014, https://www.cbinsights.com/research-reports/Boom-in-Global-Fintech-Investment.pdf.

190 *Although New York and London*: Ibid.

191 *Berman brought the intellectual*: "Farm 2050: Seeding the future of AgTech," Farm 2050, http://www.farm2050.com/#index.

193 *For example, having lost out*: Henning Kagermann, Wolfgang Wahlster, Johannes Helbig, and Acatech, "Securing the Future of German Manufacturing Industry: Final Report of the Industrie 4.0 Working Group. Recommendations for Implementing the Strategic Initiative INDUSTRIE 4.0," April 2013, http://www.acatech.de/fileadmin/user_upload/Baumstruktur_nach_Website/Acatech/root/de/Material_fuer_Sonderseiten/Industrie_4.0/Final_report__Industrie_4.0_accessible.pdf.

194 *Pasture Meter uses advanced*: "Senior Innovation Advisor for US Secretary of State Picks C-Dax Pasture Meter as the Innovation Highlight for New Zealand," Latest News, Pasture Meter, September 2012, http://www.pasturemeter.co.nz/view.php?main=news.

194 *Anyone with a phone*: "The Benefits," Pasture Meter, http://www.pasturemeter.co.nz/view.php?main=benefits.

194 *Sales of beef from New Zealand*: Chloe Ryan, "Focus on New Zealand-China syndrome," *GlobalMeatNews*, April 29, 2014, http://www.globalmeatnews.com/Analysis/Focus-on-New-Zealand-China-syndrome.

194 *China surpassed New Zealand's neighbor*: "Invest in New Zealand: Statistics," New Zealand Trade and Enterprise, https://www.nzte.govt.nz/en/invest/statistics//.

195 *He has proposed that Detroit*: Andreessen, "Turn Detroit into Drone Valley."

196 *Today 54 percent of the world's*: "World's Population Increasingly Urban with

More Than Half Living in Urban Areas," United Nations, July 10, 2014, http://
www.un.org/en/development/desa/news/population/world-urbanization-
prospects-2014.html; Parag Khanna, "Beyond City Limits," *Foreign Policy*,
August 16, 2010, http://www.foreignpolicy.com/articles/2010/08/16/beyond_
city_limits?page=0,0.

196 *Cities are incubators of growth*: Andrew F. Haughwout and Robert P.
Inman, "How Should Suburbs Help Their Central Cities? Growth and
Welfare Enhancing Intra-metropolitan Fiscal Distributions," Federal Reserve Bank of New York, 2004, http://www.newyorkfed.org/research/
economists/haughwout/suburbs_help_central_cities_haughwout.pdf.

197 *Talent can be more effectively*: Edward L. Glaeser, "Cities, Information, and
Economic Growth," *Cityscape* 1, no. 1 (1994): 9–47, http://www.huduser
.org/periodicals/cityscpe/vol1num1/ch2.pdf.

197 *Exporting advanced services*: Sir Peter Hall, "The World's Urban Systems: A
European Perspective," *Global Urban Development Magazine* 1, no. 1 (2005),
http://www.globalurban.org/Issue1PIMag05/Hall%20article.htm.

197 *Each fills a particular service*: "The A. T. Kearney Global Cities Index and Global
Cities Outlook 2015," A. T. Kearney, May 20, 2015, http://www.atkearney
.com/gbpc/global-cities-index/full-report/-/asset_publisher/yAl1OgZpc1DO/
content/2012-global-cities-index/10192.

199 *Pakistani army checkpoints*: Gohar Mehsud, "Waziristan: Tribal Residents
Caught between Drones, the Pakistani Army and Insurgents," *London Pro-
gressive Journal*, January 12, 2014, http://londonprogressivejournal.com/
article/view/1709/waziristan-tribal-residents-caught-between-drones-the-
pakistani-army-and-insurgents; Ahmed Wali Mujeeb, "Inside Pakistan's
Drone Country," *BBC News*, October 4, 2012, http://www.bbc.com/news/
world-asia-india-19714959.

199 *Many villages are ghost towns*: "US and Pak Adapt Their Approach on Di-
visive Issue of North Waziristan: WP," *Nation*, April 14, 2010, http://www
.nation.com.pk/politics/14-Apr-2010/US-and-Pak-adapt-their-approach-on-
divisive-issue-of-North-Waziristan-WP.

199 *In the most recent election*: "Pakistani Women Stopped from Voting in Waziristan," *Al-Arabiya*, May 11, 2013, http://english.alarabiya.net/en/News/
asia/2013/05/11/Pakistani-women-stopped-from-voting-in-Waziristan.html.

199 *Maria's family is from Waziristan*: Maria Umar, interview with Teal Pennebaker, January 6, 2014.

203 *He reportedly has never*: "Russia," OpenNet Initiative, December 19, 2010,
https://opennet.net/research/profiles/russia.

204 *A total of $2.5 billion*: Antonio Regalado, "In Innovation Quest, Regions
Seek Critical Mass," *MIT Technology Review*, July 1, 2013, http://www
.technologyreview.com/news/516501/in-innovation-quest-regions-seek-
critical-mass/; "Opportunities for Industrial Partners," Skolkovo Innovation
Centre, http://aebrus.ru/upload/Skolkovo%20Foundation%20AEB%2029-
01-2014.pdf.

204 *In a notable contrast*: "Women, Business and the Law 2014: Removing Restrictions to Enhance Gender Equality," International Bank for Reconstruction and Development/World Bank, 2013, http://wbl.worldbank.org/~/media/FPDKM/WBL/Documents/Reports/2014/Women-Business-and-the-Law-2014-FullReport.pdf.

205 *Industrial production dropped*: Zvi Lerman, Yoav Kislev, David Biton, and Alon Kriss, "Agricultural Output and Productivity in the Former Soviet Republics," University of Chicago, 2013, http://www2.econ.iastate.edu/classes/econ370/shuffman/documents/510410.web.pdf.

205 *Inflation skyrocketed to over 1,000 percent*: Kalle Muuli, "One Reason for Estonia's Success Lies across the Gulf," *Ukrainian Week*, May 26, 2013, http://ukrainianweek.com/World/80437.

205 *"To get my country out"*: Nathalie Vogel and Dmitry Udalov, "Who Is Afraid of Mart Laar?" World Security Network, November 12, 2005, http://www.worldsecuritynetwork.com/Europe/and-Dmitry-Udalov-Nathalie-Vogel-1/Who-is-afraid-of-Mart-Laar.

206 *In ending subsidies to state-owned companies*: Mart Laar, "The Estonian Economic Miracle," Heritage Foundation, August 7, 2007, http://www.heritage.org/research/reports/2007/08/the-estonian-economic-miracle.

206 *In 2000 Internet access was*: Colin Woodard, "Estonia, Where Being Wired Is a Human Right," *Christian Science Monitor*, July 1, 2003, http://www.csmonitor.com/2003/0701/p07s01-woeu.html.

207 *Its GDP of over $25,000 per capita*: "Country Comparison: GDP—per capita (PPP)," IndexMundi, http://www.indexmundi.com/g/r.aspx?v=67.

208 *Opponents should expect that Lukashenko*: "Profile: Alexander Lukashenko," *BBC News*, January 9, 2007, http://news.bbc.co.uk/2/hi/europe/3882843.stm.

208 *Around 40 percent of industrial enterprises*: Zuzana Brixiova, "Economic Transition in Belarus: Achievements and Challenges," International Monetary Fund, June 9, 2004, http://www.imf.org/external/country/blr/rr/pdf/060904.pdf.

208 *Belarus's currency is the Belarusian ruble*: "Belarus," in *CIA World Factbook*, https://www.cia.gov/library/publications/the-world-factbook/geos/bo.html; Leonid Bershidsky, "Russian Ruble's Hapless Little Brother," *Bloomberg View*, February 4, 2015, http://www.bloombergview.com/articles/2015-02-04/russian-ruble-s-hapless-little-brother.

209 *President Ilves is not quite like*: Toomas Ilves, interview with Alec Ross, January 13, 2014.

209 *It has the world's fastest*: Nina Kolyako, "Estonia Ranked First Worldwide in Terms of Broadband Internet Speeds," *Baltic Course*, January 27, 2012, http://www.baltic-course.com/eng/good_for_business/?doc=52217.

210 *In 2007, Estonia became*: Eric B. Schnurer, "E-Stonia and the Future of the Cyberstate," *Foreign Affairs*, January 28, 2015, http://www.foreignaffairs.com/articles/142825/eric-b-schnurer/e-stonia-and-the-future-of-the-cyberstate.

210 *Ninety-five percent of Estonians file*: A. A. K., "How Did Estonia Become a Leader in Technology?" *Economist*, July 30, 2013, http://www.economist .com/blogs/economist-explains/2013/07/economist-explains-21?zid=307&a h=5e80419d1bc9821ebe173f4f0f060a07.

210 *As the country put more*: L. S., "Not Only Skype," *Economist*, July 11, 2013, http://www.economist.com/blogs/schumpeter/2013/07/estonias-technology-cluster.

210 *The country now spends*: "Government Expenditure per Student, Primary (% of GDP per Capita)," World Bank, http://data.worldbank.org/indicator/ SE.XPD.PRIM.PC.ZS.

211 *School enrollment and literacy*: "Literacy Rate, Adult Total (% of People Ages 15 and above)," World Bank, http://data.worldbank.org/indicator/ SE.ADT.LITR.ZS.

211 *Computer language is just another*: Ilves, interview.

212 *The very name* Ukraine: Steven K. Pifer, "Ukraine or Borderland?" *New York Times*, October 28, 2011, http://www.nytimes.com/2011/10/29/ opinion/29iht-edpifer29.html?_r=0.

213 *Enable Talk took home*: "Best Inventions of the Year 2012: Enable Talk Gloves," *Time*, November 1, 2012, http://techland.time.com/2012/11/01/ best-inventions-of-the-year-2012/slide/enable-talk-gloves/.

214 *The main slogan of his*: "Profile: Ukraine's President Petro Poroshenko," *BBC News*, June 7, 2014, http://www.bbc.com/news/world-europe-26822741.

216 *There are all sorts*: "Most Innovative in the World 2014: Countries," Bloomberg Rankings, January 7, 2014, http://images.businessweek.com/bloomberg/ pdfs/most_innovative_countries_2014_011714.pdf.

216 *It places restrictions on free*: "Table: Religious Diversity Index Scores by Country," Pew Research Center, Religion and Public Life, April 4, 2014, http://www.pewforum.org/2014/04/04/religious-diversity-index-scores-by-country/?utm_content=buffer78c96&utm_medium=social&utm_ source=twitter.com&utm_campaign=buffer.

216 *Economic reforms have lifted*: "China Overview," World Bank, http://www .worldbank.org/en/country/china/overview.

218 *In the FTZ, the yuan*: Lotus Yuen, "Just How Free Is Shanghai's New Free Trade Zone?" *Foreign Policy*, October 4, 2013, http://blog.foreignpolicy.com/ posts/2013/10/04/just_how_free_is_shanghais_new_free_trade_zone.

218 *One visible example is Microsoft*: Shen Hong, "One Year On, Shanghai Free-Trade Zone Disappoints," *Wall Street Journal*, September 28, 2014, http:// www.wsj.com/articles/one-year-on-shanghai-free-trade-zone-disappoints-1411928668.

218 *No one in their rational mind*: Mei Xinyu, "China Does Not Set the 'Political Concessions,'" *Wanghai Online* (Chinese language), September 27, 2013, http://paper.people.com.cn/rmrbhwb/html/2013-09/27/content_1304366 .htm.

218 *Currently these industries comprise*: Lan Lan, "Nation Seeks Strategic

Industries' Development," *China Daily*, July 24, 2012, http://www.china-daily.com.cn/china/2012-07/24/content_15610285.htm.

219 *Its citizens speak*: People's Linguistic Survey of India, http://peopleslinguisticsurvey.org/.

219 *The World Bank ranks it*: "Ease of Doing Business Index (1 = Most Business-Friendly Regulations)," World Bank, http://data.worldbank.org/indicator/IC.BUS.EASE.XQ.

220 *It leads the world*: Yougang Chen, Stefan Matzinger, and Jonathan Woetzel, "Chinese Infrastructure: The Big Picture," *McKinsey Quarterly*, June 2013, http://www.mckinsey.com/insights/winning_in_emerging_markets/chinese_infrastructure_the_big_picture.

220 *India trains around 1.5 million*: Anumeha Chaturvedi and Rahul Sachitanand, "A Million Engineers in India Struggling to Get Placed in an Extremely Challenging Market," *Economic Times*, June 18, 2013, http://articles.economictimes.indiatimes.com/2013-06-18/news/40049243_1_engineers-iit-bombay-batch-size.

220 *However, India has neglected*: "Nehru's Approach to Primary Education Lamentable: Amartya," *Economic Times*, July 4, 2011, http://articles.economictimes.indiatimes.com/2011-07-04/news/29736088_1_primary-education-higher-education-educational-system.

222 *The biometric cards are called*: Unique Identification Authority of India, Government of India, http://uidai.gov.in.

222 *After the Modi government began*: "Downwardly Mobile," *Economist*, January 29, 2015, http://www.economist.com/news/finance-and-economics/21641272-banks-have-signed-up-120m-customers-five-months-was-easy.

222 *As of this writing*: "AADHAAR Generation Progress in India," Unique Identification Authority of India, Government of India, https://portal.uidai.gov.in/uidwebportal/dashboard.do?lc=h.

222 *In the 2000s, Brazil did*: Patti Domm, "Growing Middle Class Fuels Brazil's Economy," CNBC, April 28, 2011, http://www.cnbc.com/id/42785493#.

223 *During the same period, Argentina*: "Immigration to Argentina," *Wikipedia*, http://en.wikipedia.org/wiki/Immigration_to_Argentina#cite_note-ref1-1.

223 *By 1914, Argentina ranked*: "A Century of Decline," *Economist*, February 15, 2014, http://www.economist.com/node/21596582/print.

224 *With oil reserves set to dry*: Ambrose Evans-Pritchard, "Saudi Oil Well Dries Up," *Telegraph*, September 5, 2012, http://blogs.telegraph.co.uk/finance/ambroseevans-pritchard/100019812/saudi-oil-well-dries-up/; "Gross Domestic Product 2014," World Development Indicators database, World Bank, July 1, 2015, http://databank.worldbank.org/data/download/GDP.pdf.

224 *The gleaming new complex*: Jeffrey Mervis, "Growing Pains in the Desert," *Science*, December 7, 2012, http://twitmails3.s3-website-eu-west-1.amazonaws.com/users/325535741/89/attachment/Science_KAUST.pdf.

224 *KAUST faculty have fumed*: Susan Schmidt, "Saudi Money Shaping US Research," *National Interest*, February 11, 2013, http://nationalinterest.org/commentary/

saudi-money-shaping-us-research-8083; "Social Progress Index 2015," Social Progress Imperative, http://www.socialprogressimperative.org/data/spi; "Global Gender Gap Report 2012: The Best and Worst Countries for Women," *Huffington Post*, October 24, 2012, http://www.huffingtonpost.com/2012/10/24/global-gender-gap-report-2012-best-worst-countries-women_n_2006395.html.

225 *The day after I met Maria Umar*: Ashfaq Yusufzai, Peshawar, and Harriet Alexander, "Malala Yousafazi to Address the UN as friends in Swat Valley Listen with Pride," *Telegraph*, July 12, 2013, http://www.telegraph.co.uk/news/worldnews/asia/pakistan/10174882/Malala-Yousafazi-to-address-the-UN-as-friends-in-Swat-Valley-listen-with-pride.html; Ben Brumfield and David Simpson, "Malala Yousafzai: Accolades, Applause and a Grim Milestone," CNN, October 9, 2013, http://www.cnn.com/2013/10/09/world/asia/malala-shooting-anniversary/.

225 *It does not matter how*: Zara Jamal, "To Be a Woman in Pakistan: Six Stories of Abuse, Shame, and Survival," *Atlantic*, April 9, 2012, http://www.theatlantic.com/international/archive/2012/04/to-be-a-woman-in-pakistan-six-stories-of-abuse-shame-and-survival/255585/; "Pakistan," in *The World Factbook*, Central Intelligence Agency, https://www.cia.gov/library/publications/the-world-factbook/geos/pk.html.

226 *Countries that are closing the gender gap*: "Gender Parity," World Economic Forum Agenda, http://www.weforum.org/issues/global-gender-gap.

226 *In the developing world*: Katty Kay and Claire Shipman, "Fixing the Economy Is Women's Work," *Washington Post*, July 12, 2009, http://www.washingtonpost.com/wp-dyn/content/article/2009/07/10/AR2009071002358.html.

226 *According to the World Bank*: "Women, Business and the Law," World Bank Group, http://wbl.worldbank.org/data.

226 *The world's largest Muslim-majority*: "List of Islands of Indonesia," *Wikipedia*, http://en.wikipedia.org/wiki/List_of_islands_of_Indonesia.

227 *This culture extends to government*: Yenni Kwok, "Indonesia's Elections Feature Plenty of Women, But Respect in Short Supply," *Time*, April 8, 2014, http://time.com/53191/indonesias-election-features-plenty-of-women-but-respect-in-short-supply/.

228 *In factories they were given*: "The Lives of Rural and Urban Chinese Women under State Capitalism," Mount Holyoke College, https://www.mtholyoke.edu/~jejackso/Women%20Under%20Mao.htm.

228 *A quarter of urban women*: "Pick and Choose," *Economist*, May 1, 2014, http://www.economist.com/news/books-and-arts/21601486-why-womens-rights-china-are-regressing-pick-and-choose.

228 *In 2013, China led*: "Women in Senior Management: Setting the Stage for Growth," Grant Thornton International Business Report 2013, http://www.thebigidea.co.uk/wp-content/uploads/2014/05/Grant-Thornton.pdf.

228 *Half of the world's wealthiest*: Jonathan Kaiman, "Chinese Women Move Up Ranks of Global Super-Rich," *Guardian*, September 18, 2013, http://www.theguardian.com/business/2013/sep/18/chinese-women-global-super-rich-lists.

228 *One-third of its board*: Charles Riley, "The Women of Alibaba Put Silicon Valley to Shame," *CNN Money*, June 18, 2014, http://money.cnn.com/2014/06/18/technology/alibaba-gender-diversity/.

228 *This high-quality education*: "Holding Back Half the Nation," *Economist*, March 27, 2014, http://www.economist.com/news/briefing/21599763-womens-lowly-status-japanese-workplace-has-barely-improved-decades-and-country.

228 *As a point of comparison*: "Women in Work: The Norwegian Experience," *OECD Observer* 293, no. 4 (November 2012), http://oecdobserver.org/news/fullstory.php/aid/3898/Women_in_work:_The_Norwegian_experience.html.

228 *Japanese women comprise less*: Tomoko Otake, "Japanese Women Strive to Empower Themselves," *Japan Times*, March 3, 2013, http://www.japantimes.co.jp/life/2013/03/03/people/japanese-women-strive-to-empower-themselves/#.U5UPgJRqrlc.

228 *The numbers are no better*: "Holding Back Half the Nation."

228 *And across the board*: "Japan Remains Near Bottom of Gender Gap Ranking," *Japan Times*, October 29, 2014, http://www.japantimes.co.jp/news/2014/10/29/national/japan-remains-near-bottom-of-gender-gap-ranking/#.VPNDcbPF_pA.

229 *That translates to either*: Mark Fabian, "Japan Needs to Reform Its Work-Hour Culture," East Asia Forum, January 22, 2014, http://www.eastasiaforum.org/2014/01/22/japan-needs-to-reform-its-work-hour-culture/.

229 *In a speech at the World Economic Forum*: "A New Vision from a New Japan," World Economic Forum 2014 Annual Meeting, Speech by Prime Minister Abe, January 22, 2014, http://japan.kantei.go.jp/96_abe/statement/201401/22speech_e.html.

229 *Too often, child care centers*: "Holding Back Half the Nation."

229 *By contrast, China often relies*: Kelly Yang, "In China, It's the Grandparents Who 'Lean In,'" *Atlantic*, September 30, 2013, http://www.theatlantic.com/china/archive/2013/09/in-china-its-the-grandparents-who-lean-in/280097/.

230 *The economic importance of this fact*: Yuka Hayashi, "Japan Releases Another Plank of Abenomics Aimed at Growth," *Wall Street Journal*, June 16, 2014, http://online.wsj.com/articles/japan-releases-another-plank-of-abenomics-aimed-at-growth-1402928723.

230 *He declared that by 2020*: "Holding Back Half the Nation."

230 *"That is what Hillary Clinton*: http://japan.kantei.go.jp/96_abe/statement/201401/22speech_e.html.

231 *Going through the profiles*: "Our Team," Civis Analytics, https://civisanalytics.com/team/.

232 *By contrast, in more rigidly*: Li Qian, "Chinese Dominate Ranking of Young CEOs," *China Daily*, January 24, 2007, http://www.chinadaily.com.cn/china/2007-01/24/content_791703.htm.

232 *China's biggest social media*: "Ma Huateng," *Wikipedia*, http://en.wikipedia.org/wiki/Ma_Huateng.

232 *Its largest e-commerce company*: Michelle FlorCruz, "Who Is Jack Ma? Five

Things to Know about the Alibaba Founder before the IPO," *International Business Times*, May 6, 2014, http://www.ibtimes.com/who-jack-ma-five-things-know-about-alibaba-founder-ipo-1580890; "Lei Jun," *Wikipedia*, http://en.wikipedia.org/wiki/Lei_Jun.

235 *Grain is nicknamed "white oil" because*: Jon Gosier, "A Look at the Apps4Africa 2011 Winners," *Appfrica* (blog), January 14, 2012, http://blog.appfrica.com/2012/01/14/a-look-at-the-apps4africa-2011-winners/.

235 *Agriculture provides 85 percent of exports*: "Tanzania," in *The World Factbook*, Central Intelligence Agency, https://www.cia.gov/library/publications/the-world-factbook/geos/tz.html.

235 *To stabilize the market*: Prue Goredema, "Youthful Innovation at Apps4Africa," *eLearning Africa News*, February 2, 2012, http://www.elearning-africa.com/eLA_Newsportal/youthful-innovation-at-apps4africa/.

235 *In Kenya I was struck*: David Talbot, "African Social Networks Thrive in a Mobile Culture," *MIT Technology Review*, April 19, 2012, http://www.technologyreview.com/news/427682/african-social-networks-thrive-in-a-mobile-culture/.

236 *One farmer joked to Kahumbu*: "iCow: Tips and Tricks for Farmers via SMS," *New Africa*, March 18, 2014, http://www.thenewafrica.info/icow-tips-tricks-farmers-via-sms/.

236 *iCow also alerts farmers*: Mfonobong Nsehe, "The Best African Mobile Apps: iCow," *Forbes*, August 2, 2011, http://www.forbes.com/sites/mfonobongnsehe/2011/08/02/the-best-african-mobile-apps-icow/; Suzannah Schneider, "Five Ways Cell Phones Are Changing Agriculture in Africa," Food Tank, January 25, 2015, http://foodtank.com/news/2015/01/five-ways-cell-phones-are-changing-agriculture-in-africa.

236 *Apps4Africa matches innovative*: Schneider, "Five Ways Cell Phones Are Changing Agriculture."

236 *iCow was developed specifically*: "The Big Picture: Facts and Figures," Go Dairy, DairyNZ Limited, http://www.godairy.co.nz/the-big-picture/facts-and-figures.

237 *"Women in the private sector"*: Josh Kron, "Women Entrepreneurs Drive Growth in Africa," *New York Times*, October 10, 2012, http://www.nytimes.com/2012/10/11/world/africa/women-entrepreneurs-drive-growth-in-africa.html?pagewanted=all&_r=0.

237 *In a number of countries*: "GEM 2012 Sub-Saharan Africa Regional Report," Global Entrepreneurship Monitor, 2012, www.gemconsortium.com/report.

237 *It now connects all*: Bosco K. Hitimana, "Why Rwanda Economy Bounced Back to Strong Growth," Rwanda News Agency, June 30, 2014, http://rnanews.com/economy/8797-why-rwanda-economy-bounced-back-to-strong-growth/; Nicholas Kulish, "Rwanda Reaches for New Economic Model," *New York Times*, March 23, 2014, http://www.nytimes.com/2014/03/24/world/africa/rwanda-reaches-for-new-economic-model.html?_r=0.

238 *Unlike many other economies*: "Rwanda: Country Overview," World Bank, http://www.worldbank.org/en/country/rwanda/overview.

238 *His government placed gender*: "Rwanda: Gender Assessment: Progress towards Improving Women's Economic Status," African Development Bank Group, November 2008, http://www.afdb.org/fileadmin/uploads/afdb/Documents/ Project-and-Operations/rwanda.pdf.

238 *This move proved critical*: Kron, "Women Entrepreneurs Drive Growth"; "Rwanda: Economy," Michigan State University, globalEDGE, http://globaledge .msu.edu/countries/rwanda/economy.

238 *It turns out that Rwanda*: "Women in National Parliaments," Inter-Parliamentary Union, http://www.ipu.org/wmn-e/classif.htm.

CONCLUSION: THE MOST IMPORTANT JOB YOU WILL EVER HAVE

246 *He points to the example*: Ilves, interview.

247 *Codeacademy counts more*: Codecademy, http://www.codecademy.com/?d9 6a349c52fc4f68eea46a47ccb3d360.

247 *To date, more than 5 million*: "About Scratch," MIT, http://scratch.mit.edu/ about/.

248 *In addition to the billion people*: "The World's Billionaires," *Forbes*, http:// www.forbes.com/billionaires/list/.

INDEX

ABOUT THE AUTHOR

Alec Ross is one of America's leading experts on innovation. He served for four years as Senior Advisor for Innovation to Secretary of State Hillary Clinton, a role that earned him a Distinguished Honor Award from the State Department. He is currently a Distinguished Visiting Fellow at Johns Hopkins University and serves as an advisor to investors, corporations, and government leaders. Ross lives in Baltimore with his wife and their three young children.